D0330540

# Shooting for the Moon

**The Strange History of Human Spaceflight** | BOB BERMAN

**The Lyons Press**
Guilford, Connecticut

An imprint of The Globe Pequot Press

To buy books in quantity for corporate use
or incentives, call **(800) 962–0973**
or e-mail **premiums@GlobePequot.com**.

Copyright © 2007 Bob Berman

ALL RIGHTS RESERVED. No part of this book may be reproduced or transmitted in any form by any means, electronic or mechanical, including photocopying and recording, or by any information storage and retrieval system, except as may be expressly permitted in writing from the publisher. Requests for permission should be addressed to The Lyons Press, Attn: Rights and Permissions Department, P.O. Box 480, Guilford, CT 06437.

The Lyons Press is an imprint of The Globe Pequot Press.

10  9  8  7  6  5  4  3  2  1

Printed in the United States of America

ISBN  978-1-59921-031-5

Library of Congress Cataloging-in-Publication Data is available on file.

# Contents

# Introduction

*Okay.*

—Buzz Aldrin, July 20, 1969,
first word spoken from the moon's surface

It doesn't matter that thirty-five years have passed, that Woodstock teens are doting grandparents now. The mere fact that humans took the expensive and dangerous journey to the moon still stands as the greatest adventure of all time.

The job was performed by a bureaucracy—the National Aeronautics and Space Administration, better known as NASA—but the excitement was anything but government-issue. Life and death hung in the balance, human emotions occasionally brought astronauts to the snapping point, lives were dedicated, glorified, even sacrificed. Only in recent years has the full impact trickled out, as violent deaths in the current Space Shuttle program reveal just how risky and lucky were the Apollo missions.

Also, as the twelve who walked the moon watched their post-landing lives unfold, we had a chance to see how such an experience changes people. Moreover, ever since President George W. Bush asked NASA in January 2004 to start making plans to return to the moon, the experiences of those who went there have become increasingly instructive.

Back in 1972, at the conclusion of Apollo, the verdict on the moon seemed to be that it is too harsh; permanent habitation is pointless. Now space scientists have begun a 180-degree pivot, with actual money-out-of-pocket plans to set up eventual human bases to take advantage of the lunar polar ice we didn't previously know existed.

On the science level, we have learned what the 842 pounds of returned lunar rocks and dust told us about both our worlds, Earth and moon, and watched subsequent discoveries change scientists' thinking. But more than the science, the human tragicomedies, the astounding technology, and quirky mistakes, there is the sheer epic adventure of it all. In terms of manpower and time consumed, Project Apollo closely matched the building of the Great Pyramid. In terms of money, it was no less an expense than the Manhattan Project that produced the A-bomb. In resources, it dwarfed the building of Hoover Dam. If those enterprises can fill volumes, does Apollo deserve less?

Most previous volumes recounted astronaut biographies or dryly summarized each mission. Others focused on a single splashy mission—usually *Apollo 11* or the ill-fated *Apollo 13*. But here we will look at the entire human picture of adventure and misadventure, together with an examination of the powerful rocket that got them there. In a people-friendly way that avoids excessive technospeak but still preserves the project's meat and substance, this book will explore the dramatic, dangerous, difficult, little-known, and sometimes comical aspects of Apollo, without dumbing it down.

We'll launch our adventure with a brief history of moon-dreams in literature and film, along with the evolution of rockets and spaceflight, and the U.S. manned space program that culminated in the landings. We will also take a look at the Men of Tranquillity—the astronauts who served as our emissaries to another world. How were they selected? What made each of them the unique man he was, and is? We'll share the "forgotten" missions, complete with all their mistakes and quirky occurrences, and we'll include the development of the most awesome rocket ever built, the incredible Saturn V. We'll also save some room for a look at how the newly hatched plans to return will be different from the original Apollo flights of 1969–1972. And the controversy it is already stirring.

Through it all, there is the astonishing concept of humans leaving their planet and whizzing through space's vacuum a dozen times faster than a high-velocity rifle bullet, and the peculiar risks this naturally provokes. You and I are not likely ever to do this, so we are left with vicariously experiencing the geniuses who figured it out—and the almost recklessly brave people who actually went to a world that has puzzled humans ever since we first developed brains large enough to be tormented.

Obviously, a thorough look at the engineering steps and human performers involved in all flight above the atmosphere would require a far larger volume than the present one. Here, after looking at the moon missions, we'll not concern ourselves with the many flights that merely placed humans 250 miles up aboard later orbiters such as *Apollo Suyoz, Mir,* the Shuttles, and the International Space Station, among others. Our focus instead stays with missions to other celestial bodies. Doing so, we will perform a realistic evaluation of the present plans to return to the moon and venture onward to Mars, and draw our own conclusions about whether Armstrong's "small step" will be followed in our lifetime by others.

Let us leave our planet. It is a voyage not of one period of time, but an odyssey that will span centuries.

# CHAPTER 1

# Dreams of the Moon

*Moon!*
*Moon!*
*I am prone before you.*
*Pity me,*
*And drench me in loneliness.*

—Amy Lowell, "On a Certain Critic"

The desire to explore, to even leave our world, has always been as natural as burning the pot roast. Actually overcoming the pull of gravity is, of course, another story.

If we *did* somehow manage to escape our planet, the obvious first destination was always the moon. Even the oldest civilizations knew that the moon must be our closest celestial neighbor: it moves fastest through the sky, and in everyday life things nearby appear to whiz most rapidly through one's field of view. More important, the moon was obviously a *place*. This was no small fact. Planets—mere dots that changed brightness and moved irregularly—were widely regarded as gods rather than destinations.

So the moon beckoned. It was intriguing—even maddeningly so—to most societies. In addition to supplying much-needed nocturnal light and changing shape like a phantom in a dream, and somehow influencing the in-and-out breaths of the very seas, the moon was a lockbox of frustrating paradoxes. For thousands of years, long before Galileo aimed the first telescope at the moon on January 7, 1610, bizarre moon qualities tormented the minds of the greatest thinkers.

Whenever a sphere is illuminated straight on by a light source, its center appears noticeably brighter than the edges. It's a common experience, and the main reason a ball appears three-dimensional, even in a painting. But the ancient Greeks noticed that the full moon completely lacks this dimensionality. Instead it's evenly lit—as two-dimensional as a dinner plate.

The full moon seems like a disk *painted* onto the sky. Until the present era, no one could begin to guess why.

Then there was its odd light. You'd think that the full moon would be twice as bright as a half moon. But it's not even close. The full moon is ten times brighter. The ancients noticed this, but had no explanation. Nor could they explain the "moon illusion"—the fact that the moon looks huge when it's low down, near the horizon. But perhaps of equal mystical import was the moon's width. Their sky displayed thousands of stars and planets, all of which were points of light. But the heavens possessed only two disks, sun and moon. Both appeared *exactly* the same size! Moreover, this perfect match occasionally allowed the moon to fit precisely over the sun to block it out and yield the stupefying experience of the total solar eclipse. Could it *be* more wondrous?

Then, too, there was the fact that looking up at a full moon meant gazing at perfect roundness. Its four-mile equatorial bulge is utterly indiscernible against its 2,160-mile width. Visually, it's a perfect sphere. While the sun, too, appears round to the naked eye, its brilliance makes it difficult to see except when viewed through thick horizon air at sunrise or sunset, and at those times the atmosphere distorts it. So with rare exceptions, one views nature's most flawless circle only during a full moon. Considering the circle's ancient and pervasive connection with perfection, its societal associations (such as wedding rings), and its wondrous mathematical qualities, the exquisite roundness of the full moon was one more way in which it inspired its classical sense of poetry, romance, and mystery.

All this was part of life. But actually making a journey to see the moon firsthand? That was too preposterous other than as an occasional fantasy-myth of the ancient Greeks and a few others. Even then, such notions always had a tongue-in-cheek quality. Written between the lines was forever the implied assumption: no one is ever actually going to do this.

That is, until the age of cannon and explosives.

The Chinese, who invented gunpowder from a mixture of three common chemicals (nitrate of potassium, charcoal, and sulfur), developed rockets at around the same time the Roman Empire was building its marvelous aqueducts. Gunpowder's use in weaponry was well established by the thirteenth century, when the Chinese possessed artillery augmented by rockets. By the year 1494, heavy immobile cannons—now firmly established in Europe—were being replaced with mobile units that brought the castle-busting armaments to the view of many increasingly alarmed people. After John Maritz invented a machine tool that could bore cannon barrels in 1713, refining the weapons and making them more uniform, scientists of the time started wondering if continual improvements might extend their range indefinitely.

Few dreamed of firing a projectile beyond Earth until the late eighteenth century. Many credit Jules Verne's ahead-of-his-time imagination to have provided the first true stepping-stone on the highway to space in the mid-nineteenth century. His 1865 novel, *From the Earth to the Moon,* was perhaps the first great conceptual milestone of the pre-Apollo era.

Scientifically literate, Verne had his characters, led by Dr. Barbicane of the Baltimore Gun Club, embark upon a rational step-by-step approach to reach the moon using an enormous cannon. Employing realistic figures for the necessary cannon size and the explosive charge needed, and using known scientific figures for the actual escape velocity of Earth, Verne's fictitious club members constructed a partially buried super-gun that fired a projectile that gained all of its speed at the beginning, like today's space probes. Its name—closely resembling what later became the first Space Shuttle—was Columbiad. Verne's characters even sited their super-cannon in Florida where Earth's faster rotation would add to the projectile's velocity. This was exactly the same reasoning that led to the Space Center construction at Cape Canaveral a century later.

The only thing Verne ignored, possibly because he had no choice but to look the other way, was the impossible g-forces such a method would produce. One could indeed fire a cannonball at the moon, and hit it, too, except for two small problems. First, the speed the projectile would need to attain when leaving the muzzle—about ten miles per second or 36,000 miles per hour—meant that occupants accelerated to that velocity in a fraction of a second would be disheartened to find that all their internal organs had turned to jelly.

Second, the speed is fast enough to produce intense frictional heating with the atmosphere. Much of the projectile would be melted in the same manner that incoming meteors are—if it even survived the blast at all. (Were it not for the atmosphere slowing down the projectile during its first eight seconds of flight, the initial velocity need only be seven miles per second, or 25,000 miles per hour.)

Still, Verne was the first to make such a project *seem* feasible, and to catch the public fancy. This was more than a small accomplishment: implanting a dream or goal in the collective public mind is usually a first critical step toward its realization.

The Victorian era then produced a series of popular books that built on the space-travel theme, culminating perhaps in H. G. Wells's famous 1901 novel, *The First Men in the Moon*, which was serialized and widely read. These inspired a silent motion picture, Georges Melies's *A Trip to the Moon*, released in 1902.

Even today, more than a century later, few have not seen clips from that smash hit, which showed the rocket, a giant artillery shell, lodged in the eye of the moon's grimacing face. The fourteen-minute silent movie even had the adventurers returning to Earth by landing prophetically in the ocean.

Decades earlier, in an excellent example of how influential a single book can be—how a solitary volume can alter world history—the Verne novel of 1865 was savored over and over by a Russian nine-year-old named Konstantin Tsiolkovsky, born in a village south of Moscow on September 17, 1857. His was a self-taught childhood, and he later admitted, "Besides books I had no other teachers."

Tsiolkovsky was a brilliant teenager, who, like many youths everywhere, soon moved to the big city (in this case Moscow), got swept up in the heady science and culture of his time, and then returned to his native Izhevskoe where he easily passed the exam to become a teacher. It was here that, far past his youth, at age forty-six, he wrote the ground-breaking book *Exploring Space with Reactive Devices*. The volume was not just accurate, it was filled with prescient mathematical concepts for reaching the moon through rocketry. It even offered designs for liquid-fuel rockets, and suggested the use of kerosene, the same fuel eventually utilized by the actual Saturn first-stage engines.

Although Tsiolkovsky was ahead of his time, he was neither a designer nor a builder. It took others to take his ideas and run with them.

The first was an American, the shy and almost secretive Robert Goddard, who created his initial gunpowder rockets in Worcester, Massachusetts, in 1911. With a doctorate in physics, Goddard was more than a mere tinkerer; he was granted two patents in 1914 for rocket propulsion systems, and received a $5,000 grant from the Smithsonian Institution—a fortune at the time—to fund research into rocket fuel. In 1920, his first and best work was published, *A Method of Reaching Extreme Altitudes*, in which he, as did Tsiolkovsky, advocated the use of liquid fuel.

Shunning publicity, Goddard's real breakthrough design happened on March 16, 1926, with the first successful test of his liquid-fueled engine, launched from a relative's Massachusetts farm. Using the hard-to-produce liquid oxygen mated with gasoline, a more dangerous brew than the oxygen-kerosene that would later launch most of the world's space satellites even in the twenty-first century, the Goddard rocket only achieved an altitude of forty-one feet—no taller than a four-story building—but it was nonetheless a milestone. Unfortunately for his country, it would be others in Europe who would create all further major advances in rocketry for the next quarter-century.

Indeed, three years earlier than Goddard's first liquid-fueled engine test, the Romanian-born physicist Hermann Oberth had published his *The Rocket into Interplanetary Space*, which helped fire Europe's widespread fascination with the topic. After reading it, famed German moviemaker Fritz Lang collaborated with Oberth to create *Woman on the Moon* (Frau im Mond), a movie that captivated audiences throughout the continent. Small wonder. The film was more than visionary; its special effects were many years ahead of their time, even depicting water droplets in the space-faring ship forming floating spherical blobs just as they eventually did in reality, decades in the future. Scene after scene in this 1929 movie appear astonishingly prescient today, as they accurately portray images that actually unfolded in the future space age. We see a huge rocket, constructed in a gargantuan assembly building, slowly wheeled out vertically to a launch station on a platform riding on rails. Before launch, a man solemnly prepares for the ignition with a backwards count from ten to one—the first time such a countdown had ever been made in fiction or in real life.

Since our story has now taken us to movie houses in pre–World War II Germany, it will be obvious where it must go next, and what person comes into focus. No single human in history is more associated with rocketry than Wehrner von Braun.

An engineering student and member of a rocketry club, von Braun became Oberth's assistant in Berlin. The fun and games ended (and, in a way, began) for these aristocratic youths in 1932 when the German military asked Oberth and von Braun to work on constructing rockets that might have more than a mere space-faring purpose. Von Braun agreed, reportedly with reluctance. Yet the handsome patrician youth apparently knew how to win friends and influence people: by the late 1930s the new secret facility located at Peenemunde was entirely under his directorship and had a staff of over 2,000. It would eventually top 10,000, if one includes the slave laborers.

By the early 1940s, the team had created by far the most colossal and powerful rocket ever made, which they called the A-4. It could hurl a ton of explosives 200 miles, at speeds faster than sound, making it almost impossible to shoot down. In late 1943, Hitler ordered the construction of thousands of these weapons, and changed the name to Vengeance Two, commonly known in abbreviated form as the V-2.

To von Braun's credit, he dragged his feet on the military side of things and kept his eyes so focused on the eventual hoped-for space travel that he was arrested by the SS in 1944 and spent two weeks in prison. Only the intervention of higher-ups won his release.

The evil and insane führer's days were, of course, numbered, and von Braun made plans to escape Germany with his engineering staff as soon as the handwriting was on the wall. Above all, they knew that they must not let themselves be captured by the soon-to-be-invading Soviets. It was a sweetheart deal that the United States made, an operation named Paperclip that rendezvoused with, and whisked to America, more than a hundred of the German scientists, along with a trainload of their equipment that even included complete V-2 rockets.

The rest, as they say, is history. It was von Braun who spearheaded the United States' far-behind efforts at rocketry in the late 1940s and 1950s. Along with an ever-growing staff of American engineers, they improved the V-2 and superseded it in the mid-1950s with a series of larger and more innovative designs. The fruit of von Braun's work gained a fearsome reality in the form of 12,000 nuclear-capped missiles with transcontinental range.

Of course, during this cold war the Soviets were not sitting idly by. Although they also had their own cadre of captured German rocket engineers, they possessed an even greater asset, one that was homegrown. His name was Sergei Korolev, and the machines he and his team created were lightyears ahead of even von Braun. Korolev had suffered the typical slings and arrows routinely meted out by Stalin's paranoid rule, and had even been imprisoned in a gulag until the authorities came to their senses and realized that they needed his engineering brilliance. It was Korolev who truly opened the space age in 1957 with mankind's first-ever Earth satellite.

The launch of *Sputnik* on October 4, 1957, stunned the world. People everywhere responded to the published schedule of when the eighteen-inch sphere would pass over their homes: crowds gathered in the streets and on rooftops, to the great discomfiture of U.S. officials who could only put on a game face and send the Soviets their congratulations. (Actually, few observers realized, then or now, that they were *not* seeing the tiny sphere of *Sputnik* itself, which was too small to be visible to the naked eye. They instead were gazing at the giant third-stage booster rocket that had placed the satellite in orbit, which circled the globe alongside it the whole time.)

The Soviets' tension-filled lead in the Space Race was to last for the next nine years, but it served a valuable purpose. It proved instrumental in ushering in the most dramatic scientific race in human history.

○

# CHAPTER 2

# Kennedy's Plan: No Choice

*They would make me believe that the Moon*
*was made of greene cheese.*

—John Firth

How does a nation decide to spend enormous sums of taxpayer money in order to go to the Moon? Can you get a hundred dollars from every citizen, just to do a science project?

The question is more than a historical musing, for its answer will ultimately determine whether or not any nation returns there in the twenty-first century, or sends manned missions to Mars and beyond.

Obviously, there must be a widespread consensus for such a fabulous, expensive undertaking. It is not enough to have the technological ability and financial resources to make something happen; there must also be an impetus among leaders and citizens to actually do it. As an example of a failed megaproject due to lack of widespread support, consider the U.S. Superconducting Super Collider Project of the 1980s. The idea was to create a fifty-mile subterranean loop in Texas, a state-of-the-art particle accelerator that would facilitate top-level research into the universe of subatomic physics. The price tag was to have been over $20 billion. The public, however, knows little about the subject, and there was no widespread clamor or appeal to go forward with this groundbreaking science. The project was canceled, leaving Europe in the forefront of cutting-edge particle research. Moreover, while many in the science community bemoaned the cancellation, little or no outcry erupted in the American press or in the halls of Congress.

A proposed mission to the moon may well have met a similar fate had there not been widespread support. But how does such backing become a reality in a citizenry that is generally ill-informed about science? Clearly, the story of the Apollo triumph must have had its roots in a massive collective desire to venture into space.

Some believe that people are inherently hard-wired to explore, that a quest to leave Earth is an almost instinctual craving. They might argue that children naturally explore their rooms, then their attics or basements, and then their neighborhoods. As soon as they are old enough, they ride bicycles to increasing distances, and later, if financially able, travel to other states and provinces, or even go abroad. Exploring new places, by this reasoning, is so natural and instinctive that a trip to the moon is an easy "sell."

Others might argue that in the case of the Apollo Project, fear and competition were instead the critical motivators. When the Soviets launched their inaugural 183-pound *Sputnik* satellite in October 1957, followed by a half-ton version just a few months later, Americans were shocked and dismayed: a rocket that could go into space could also carry nuclear bombs to their city.

When the Soviets soon sent an animal, and then, in 1961, the cosmonaut Yuri Gagarin into orbit, the public outcry to catch up to the Russians became a roar. President John F. Kennedy had no choice but to figure out what, if any, prestigious space enterprise might be achievable ahead of our cold war enemy, and commit resources to it knowing that he would have the full support of the citizenry. By this reasoning, his "go to the moon before the decade is out" speech didn't necessarily mark him as a visionary leader, but merely a follower, a responder to events and the existing public demand for action.

But even if all this is true, why the moon? Why not race the Russians for a cure for cancer, or to eradicate poverty, or some other major scientific goal? Here the answer was obvious and multifaceted. The Russians were already heading quickly into space. Space was the "in" place, the challenge, and the moon was the only logical manned destination.

Fine—but why had the Russians embarked on this quest in the first place?

Part of the answer clearly lay in the fearsome German V-2 rockets of World War II, and the nuclear-tipped ballistic missiles that were developed soon afterward by both the United States and Russia. The rockets, though strictly military, were providing the technology for space travel and it was an easy next step to employ them in that capacity.

Yet, still, there was an additional impetus. In the last chapter we saw how Jules Verne, then H. G. Wells, and then the moviemaker Fritz Lang made the idea of space travel wildly popular. From the 1930s onward, an unending series of articles in popular magazines periodically trumpeted the emerging science that would "someday" or even "soon" make a trip to the moon a reality. The public was very aware of this exciting possibility, and it had been continually romanticized and glorified. Moreover, it followed each country's self-image as a technologically advanced, forward-thinking scientific land that should take the lead in such a venture.

In the United States, this was helped along by a series of articles in *Collier's* and *Popular Science* magazines between 1952 and 1955, written by the winsome German-morphed-into-American rocket hero Wehrner von Braun. Indeed, von Braun's 1952 book, *Across the Space Frontier*, followed by a popular American film *Riders to the Stars*, made space travel seem more than compelling; it was presented as almost a fait accompli. Then in 1955, Walt Disney devoted several episodes of his popular weekly TV show to space travel. The public's fascination was now fully engaged.

But funds to actually begin planning manned space travel were not forthcoming. President Dwight D. Eisenhower had no interest whatsoever in manned space travel. He did not allocate a dime toward it, and barely funded slow-paced research aimed at creating an artificial satellite, which was well within von Braun's team's immediate grasp by the mid- to late-1950s.

If the United States had been alone in the world it is doubtful that any project aimed at the moon would ever have been approved, despite the romantic public mindset that science could accomplish such an undertaking. It took that collective American shock of the Soviet achievement to finally ignite the impetus toward urgent development of a manned program.

As for the moon mission, it was evolutionary rather than revolutionary. And, more than that, it was the obvious goal. The Russians had already beaten the West to the first satellite, the first man in space, the first woman in space, the first spacewalk, the first spacecraft to fly past the moon, the first robotic landing on the moon, the first two-man crew, the first pictures of the moon's far side, and the first soft landing on the lunar surface. In most areas, there simply wasn't any competition left. There was no longer any major goal in which to compete except for an actual manned landing on the moon's surface. If the United States could pull that off, it could in one bold stroke show the world that we were not second best in space after all.

This then, was why Project Apollo was born. There was simply no choice. We *had* to go to the moon. Everyone knew it. And everyone wanted it.

# Yes, but How?

Once President Kennedy had declared the U.S. intentions of sending men to the moon, the immediate issue became: How? Tsiolkovsky had worked out the basic math of a moon mission, Goddard had proven that liquid fuel was the way to go, and von Braun had created enormous rockets. Those were the stepping-stones. But no existing rocket had anything approaching the required power to do the job. The mighty V-2 had eventually evolved into the eighty-three-foot tall U.S. Redstone rocket, which was barely good enough to launch a lightweight satellite 160 miles up, where it would scrape the thin upper atmosphere and return like a meteor in less than a year. It could send a toaster into orbit, but struggled to lift even a single astronaut to a suborbital height. The new rocket would have to be at least forty times more powerful.

Three approaches might work, and they were all very different from one another. At the time of Kennedy's announcement, most people assumed that we'd reach the moon the same way we reached Earth orbit—with one mighty rocket making the entire round-trip. Designers called this huge hypothetical rocket the "tailsitter" because that's just what it would do: ease itself backward, gently upon the moon's surface, and then refire for the journey home.

The tailsitter concept, also known as the direct approach, required an almost unimaginably gargantuan rocket that was even given a name: Nova. But the time-pressure of designing such a colossus from scratch and still getting people on the moon within eight or nine years was more than difficult—it might not be doable. So, scratch the one-mighty-launch concept and on to plan B.

This, simply, was to launch a less-powerful tailsitter-type rocket with only enough fuel to reach Earth orbit, send up a series of other rockets laden with fuel, and gas up the monster's empty tanks in Earth orbit. We'd build a friendly orbiting service station. Such a near-Earth gathering of rockets and tankers and astronauts to do the work would take intricate planning, but it was NASA's preferred option. Not coincidentally, Wehrner von Braun favored this "Earth-orbit rendezvous" approach, too. It probably would have worked. Calculations showed that the effort would require the tailsitter to blast off from Earth with twelve million pounds of thrust, necessitating the first stage to sport eight of the newly proposed F-1 superengines.

But there was an even better idea, one that nearly didn't get off the ground so to speak.

It was a plan first formulated back in 1916 by a contemporary of Tsiokolvsky, another self-educated Russian—Yuri Kondratyuk. Few know his name today, yet it was his brainchild that a rocket could contain two separate dissimilar sections, each with its own engines. Once in the moon's vicinity, one would detach from the other and descend to the surface while the other kept orbiting the moon, waiting for its partner's return. Once back, the segments would temporarily reconnect and the space-farers would reenter the main module to fire its engines and return to Earth, leaving the other now-discarded unit behind.

The beauty of this plan was that it involved an absolute minimum weight. And the module that was to descend to the moon could be designed specifically for the task. Operating solely in the vacuum of space, it wouldn't have to be streamlined or air-resistant. There would be no wasted fuel, either. Each unneeded section could be abandoned like an empty beer can when its job had been completed. This whole scheme would be called lunar-orbit rendezvous, abbreviated forever within young NASA's evolving language of bureauspeak as LOR.

As history now shows, this is indeed the way the United States managed to send people to the lunar surface. But it almost didn't happen. Probably the real hero was an unknown NASA engineer named John Houbolt. During the early sixties when NASA was actively interested in examining all ideas to accomplish Kennedy's goal, Houbolt and several others worked out the details to Kondratyuk's old lunar-orbit rendezvous scheme and discovered that it offered the lowest weight and highest success probability, even if such a rendezvous a quarter million miles from Earth would be daring and risky.

But Houbolt was shot down. The Earth-orbit rendezvous (EOR) mode already had a life of its own, it had an existing momentum, and even in fledgling bureaucracies such as NASA this can become as immovable as China's Great Wall. NASA's heaviest hitters were already championing it, including the brilliant, legendary Mercury spacecraft designer Maxime Faget.

Frustrated, Houbolt bravely went over the heads of superiors and contacted a NASA decision-maker who was nonetheless a few rungs of the ladder from the top, who listened with an open mind. What followed were a series of heated meetings and presentations, with Houbert suffering several strong personal criticisms. In the end, however, the unambiguously superior ace card of lessened fuel and weight requirements proved hard to trump. Unlike "soft" sciences such as sociology, the "hard" sciences of mathematics and physics rely on incontrovertible principles. Mathematicians and physicists can demonstrate to other experts that mathematical figures and physical laws produce one result and not another, and no one at a conference table can dispute them; they are not subject to interpretation. The LOR scheme was simply superior. Faget, who believed even forty years later that its advocates painted too rosy a picture of its benefits, had to admit that the LOR numbers came out better than EOR. If it could work, it would be the most efficient and lightest-weight way to achieve the goal.

On July 11, 1962, the three top NASA administrators made the surprise announcement of the unexpected technique. The United States would sink or swim using lunar-orbit rendezvous.

## Why a Countdown?

The German depression-era filmmaker Fritz Lang may be more famous for his groundbreaking movie *Metropolis,* but his *Woman on the Moon* gave the world its first countdown to a rocket launch.

Although he had technical assistance from contemporary rocket-builder Oberoth, Lang probably didn't truly believe that a backwards count from ten to one was really needed prior to launching a rocket. Rather, it was a theatrical device, a dramatic way to build up suspense before his special-effects monster-rocket would take to the sky.

So why do such a countdown in real life?

Lang may not have realized it, but an actual modern rocket launch involves so many systems and subsystems that must be switched on in the correct order, that only a precise sequencing can accomplish the task. For example, the mighty Saturn V moon rocket had to have its internal computer activated six seconds before launch, pumps furiously sending fuel to the engine nozzles two seconds later, and sparks applied to the eighteen-foot nozzles one further second later still. A countdown that begins hours prior to the moment of launch is the only logical way to make sure that everything gets done at the right time and in the correct order.

# CHAPTER 3

# The Saturn V

*It is the very error of the moon;*
*She comes more near the Earth than she was wont,*
*And makes men mad.*

—William Shakespeare, *Othello*

From the very beginning, NASA engineers knew that they would have to design and build the most powerful rocket on the face of the Earth. Rockets of the late 1950s and 1960s included the 83-foot Redstone, which could not quite lift any manned payload into orbit, the 95-foot Atlas, which could hurl 3,000 pounds into orbit, and the 109-foot Titan, which could place 7,500 pounds into Earth orbit. Mere makeovers to soup them up would not suffice. It was not enough to strap on extra engines, or make them bigger. The next rocket would have to be a brand-new machine, created from scratch.

It was obvious from the beginning that the standard three-stage paradigm did not have to be chucked; it merely required a vastly greater scale.

Wehrner von Braun's team started working on the new monster's first stage using tried-and-tested fuels for reliability. After initial development work, this behemoth stage, named the S-1C in a rare moment of restraint given every bureaucracy's need to affix as many letters and numbers as possible to everything, was taken over and manufactured by the Boeing Company. The next two stages, designated S-II and S-IVB (don't ask), were contracted out to North American Aviation and McDonnell Douglas respectively, with the Rocketdyne company building the actual engines.

These engines, particularly the main power plants of the colossal S-IC lower stage, were clearly the biggest technological hurdle of the entire enterprise. From the very beginning of development, the first stage engines were to be called F-1; the air force had been working on this superengine concept for a couple of years even before the Soviets had launched *Sputnik*.

They would now have to be brought from the drawing board to reality, with their myriad problems finally solved. Each was required to produce 1.5 million pounds of thrust, which was more power than any previous entire rocket.

## Thrust!

Rocket power is often expressed as "pounds of thrust"—which is easy to understand. With airplanes, lift is supplied by air moving across the wing, creating a partial vacuum on the upper surface that literally sucks the plane upward. A propeller or jet engine need only supply enough thrust to move the plane forward fast enough so that enough high-speed air traverses the wings. With rockets, the entire lift against the force of gravity is supplied solely by the engines. If a rocket weighs a million pounds, then a force of fast-moving gas must be able to push backward with a million pounds of thrust just to suspend the rocket stationary against the pull of gravity. To move upward, excess thrust above the rocket's weight is needed. The amount of additional available thrust beyond the weight of the rocket determines how fast it will rise. In the case of Saturn V, the total first-stage power of 7.6 million pounds of thrust easily overcame the rocket's 6-million-pound blast-off weight, with 1.6 million pounds of excess force at liftoff to create upward motion. Since the rocket's astonishing fuel consumption made the Saturn 180,000 pounds lighter each minute—and atmospheric drag decreased as the craft passed quickly through the lower thick atmosphere—increasing amounts of thrust became available for acceleration as the launch proceeded.

The F-1 engine, designed and built by Rocketdyne, was a thing of beauty—the most important single piece of equipment in the moon race. By comparison the Soviets never did create a power plant of comparable size, and instead placed their moon hopes on a super-rocket called the N-1, which had an almost ludicrously unwieldy first stage, employing a crowded suite of thirty smaller engines. The failure to fire this confused horde in unison, and reliably, was what doomed the Russians. All four tests of this Soviet monster-rocket resulted in explosions.

There was some initial debate over how many F-1 engines should compose the Saturn's first stage. As noted earlier, if EOR (the initially favored Earth-orbit rendezvous method of reaching the moon) was chosen, eight engines and twelve million pounds of thrust would be required. But LOR—rendezvousing in moon orbit and sending a separate craft that itself had two detachable engines for descent and later ascent—only needed four. Thus, for nearly two years the plan called for a quad-configuration that would have supplied six million total pounds of thrust. After much back-and-forth discussion, it was decided that the proposed arrangement of four nozzles ringing the bottom allowed physical room for a fifth engine right in the center, and if a another one could physically fit, it might as well be there. This "Saturn V" concept appealed to engineers. It would provide a bit of a safety margin in the case of an engine failure; it would allow the Saturn to continue traveling farther after launch and might facilitate an escape in a situation where none would have been possible with just four. More important, it allowed for extra payload should the need later arise.

That fifth engine ultimately proved a critical blessing. Without it, there would have been insufficient thrust to hoist up a 455-pound lunar rover for driving excursions along the lunar terrain. Without it, the later use of the Saturn for launching the Skylab space station would have been impossible, and the Apollo missions themselves would have operated at their very limits.

The F-1 development had been continually fraught with setbacks. The engine had the bad habit of shaking itself to destruction; it seemed to be just too powerful. And the extreme 4,000-degree temperatures from the exploding fuel mixture was enough to melt steel, let alone the more fragile components of the combustion chamber. Consider: Each engine stood almost two stories tall, and consumed 3,000 pounds of fuel per second. This propellant, a finely sprayed mixture of kerosene and liquid oxygen, had to burn so smoothly that "burps" or momentary explosions would be unable to extinguish the inferno and shut down the engine. This was solved by forcing the fuel at high pressure through thousands of tiny holes, like the pores in a fine sieve. The melt-the-nozzle problem was solved by routing the supercold liquid oxygen fuel around and around the outside of the engine before it entered to be consumed.

Merely delivering the fuel to the engines was a vast technical headache. Though the engines were at the bottom and the fuel higher up, a simple gravity-feed system such as that used in high-wing airplanes would be inadequate. The solution was a fuel pump that was itself 65,000 horsepower. When one considers that most family automobiles have about 160 horsepower—and that's for the entire engine—a mere fuel pump with 65,000 horsepower is almost beyond imagination for its pure power.

The entire suite of five F-1 engines with its 7.5 million pounds of thrust was designed to live just two minutes, forty seconds. When the Saturn was forty-two miles high and traveling 9,000 miles per hour, it would shut down and fall back into the ocean as junk.

After a metal connector was then released—the staging ring whose slow fall-back to Earth has been spectacularly shown many times in films—the second stage would fire. This S-II stage also had a constellation of five engines, but these were much smaller power plants than the F-1s. Called the J-2s, they would be brand-new state-of-the-art designs. These were the engines the astronauts most feared: their technology and reliability had been untested by the trial and error of earlier rocket generations.

The J-2s were cryogenic, ultracold; one component was the familiar LOX of liquid oxygen, but the other was the even colder liquid hydrogen. The two elements explosively combine and, weight-for-weight, provide more thrust for their mass, more bang for the buck, than any other practical fuel. (Its success in Saturn eventually made it the sole propellant for the Space Shuttle's three main engines some fifteen years later).

The five J-2 engines together created just over one million pounds of thrust. They carried the craft to 117 miles up, the edges of space, where the third stage powered by a single J-2 would take over. After turning off to provide a few circuits around Earth to check out all systems, this single J-2 would refire to propel *Apollo* toward the moon.

Altogether, the Saturn V could place an incredible 125 tons into Earth orbit, or bring 47.5 to the moon. It would weigh 6 million pounds, and stand 363 feet tall, the height of a 36-story skyscraper.

Sounds simple perhaps. But the Saturn V had more than a million separate parts. It was the most complex flying machine of all time. It took 400,000 people laboring for nearly a decade to produce it—roughly the same man-hours that was used to build the Great Pyramid at Giza.

As an interim step, NASA also created the two-stage Saturn 1B rocket, much smaller than the full Saturn V, but capable of launching heavy satellites and probes, and placing Apollo mock-ups into orbit for testing until the full Saturn V was ready. The Saturn 1B stood only two-thirds as tall as the Saturn V, and could carry only one-sixth the amount of weight into Earth orbit.

Testing of Saturn engines took place in a special facility in rural Alabama, where the bone-rattling vibrations and decibel level demonstrated to everyone living within twenty-five miles the true meaning of "bad neighbor." The sound was nothing less than deafening, if blessedly brief. Windows would break many miles away, particularly when tests were conducted on cloudy days, when the vibrations reflect back down to the ground.

A central issue was, of course, reliability. The engineers tackled this by ensuring that each subsystem (fuel valves, electrical switches, fuel pumps, navigation electronics, and the like) had a failure rate so low (typically .9998) that when all systems' failure probabilities were multiplied together, the final result for the entire rocket would not be worse than .990. In other words, a Saturn V would only have one chance in a hundred of failing. To achieve this, backups and redundancies were built in wherever necessary.

Over the strong objections of von Braun's team, Associate Administrator for Manned Space Flight George Mueller boldly decided that time constraints did not permit separate in-flight tests of the Saturn's various stages. Thus it happened, for the first time in U.S. history, that a major new rocket was blasted off with all untested stages attempting to operate together the very first time as an ensemble. The date, November 9, 1967, goes down as a milestone. Had that first complete Saturn V seriously malfunctioned, the decade would have ended with the moon still a mere dream suspended in the sky.

○

# CHAPTER 4

# Who Will Volunteer to Live on the Moon?

*The moon is a white strange world. . . . in the night sky, and what she actually communicates to me across space I shall never fully know. But the moon that pulls the tides, and the moon that controls the menstrual periods of women, and the moon that touches the lunatics, she is not the mere dead lump of the astronomist. . . . When we describe the moon as dead, we are describing the deadness in ourselves.*

—D. H. Lawrence

The story of the U.S. manned space effort can be traced back years or decades, depending on which milestone moment, or inspirational visionary, is chosen to represent the moment of genesis. But as we've seen, pinpointing an exact date is usually no more difficult than picking October 4, 1957, the launch of the first Soviet Sputnik satellite. The public clamor and U.S. government reaction set in motion a series of events that was to take us to the moon less than a dozen years later.

By order of President Eisenhower, who had initially been extremely slow in initiating U.S. manned space flight, the plodding underfunded military space program, along with the old National Advisory Committee for Aeronautics (NACA) which had been around since World War II, was to be changed over to a new civilian agency. Thus it was that the National Aeronautics and Space Administration opened its doors on October 1, 1958, inheriting personnel and resources that included the Langley Aeronautical Laboratory, the Lewis Flight Propulsion

Laboratory, the Ames Aeronautical Laboratory, and about 8,000 engineers, scientists, technicians, and numerous supporting clerks and civil service bureaucrats.

In January of the same year, just three months after *Sputnik 1*, a three-day military meeting at Ohio's Wright-Patterson Air Force Base examined various designs for a U.S. manned capsule to space. Eleven aerospace companies attended the conference; some of their engineering offerings were spherical, some conical, some were two-man designs, two even had little wings so that they could be "flown" down from orbit. It was McDonnell Douglas that suggested a one-ton truncated cone that could be launched on a modified Atlas ballistic missile.

The main design limitation was the g-forces a human is capable of withstanding. Studies had already shown that during the processes of launch and reentry, nine g's might well be encountered, and perhaps briefly as many as sixteen. Would a man be able to survive that?

Not if he were standing up or sitting normally. But a series of tests showed that the shift of blood and internal organs could be tolerable in a specially designed couch, with the astronaut facing backward upon return.

In short order the design was approved, with a plan calling for a manned launch by April 1960. In other words, a mere one-and-a-half years was considered enough time between project approval and an actual manned orbital flight. (Way too optimistic, it turned out: in reality, the first American wasn't launched into orbit until four years later, in early 1962.)

One of the lesser details that had yet to be settled was who, exactly, would fly these dangerous, modified intercontinental ballistic missiles. As had the rest of the program, the personnel selection job fell to the newly formed Space Task Group (STG), headed by Bob Gilruth. Gilruth very quickly became one of the top players in the early space program, and pushed manned flight when there still were many hurdles, including the reluctance of President Eisenhower himself.

Dr. Robert Rowe Gilruth helped organize the Manned Space Craft Center (later called the Johnson Space Center) and was director of it from 1961 through 1973, which spanned the entire years when NASA was solely directed at sending humans to the moon.

Gilruth had come around to the idea of manned spaceflight slowly and reluctantly in the late 1950s ("I had thought it was terribly dangerous") and, as did many, recognized the need for a U.S. effort only after the first two Soviet satellites were launched. Specifically, he later recalled that it was *Sputnik 2* in November 1957, which carried a dog, Laika, on board, that really did the trick for him. When that Russian dog went up, ". . . I said to myself, and to my colleagues, 'This means that the Soviets are planning to fly a man in space.'"

But the pace really didn't quicken until John F. Kennedy took office. Shortly after his inauguration, Kennedy personally said to Gilruth, "I want to be first. Now do something." Gilruth recalls laying it bluntly on the line. He told the young president:

*Well, you've got to pick a job that's so difficult that it's new, that [the Russians] will have to start from scratch. They just can't take their old rocket and put some new gimmick on it and do something we can't do. It's got to be something that requires a great big rocket, like going to the moon. Going to the moon will take new rockets, new technology, and if you want to do that, I think our country could probably win because we'd both have to start from scratch.*

At that point the Mercury manned orbital flights were already firmly in the works, but they were a dead-end project. They really didn't lead to any next step all by themselves. President Eisenhower had cut nearly one-quarter of the space program's preliminary 1962 budget, and Kennedy had not allocated any funds for manned flight beyond the Mercury missions, although some money remained for rocket development. Existing plans stated that nothing would happen until after the completion of Mercury, when there would be further "study" about what, if anything, would be done next in the field of manned spaceflight. In bureaucratese the word *study* normally translates to: "No actual further activity." This was the gloomy background against which two pivotal events were to change everything.

First, Yuri Gagarin went up on April 12, 1961, and the Russians immediately boasted about it at every possible opportunity. Kennedy was seriously concerned that international opinion would come to unduly admire the Soviet Union. Then a few weeks later, when Alan Shepard completed his brief fifteen-minute suborbital Mercury flight on May 6, and Kennedy saw what kind of wild press and public enthusiasm he received, the president knew how to read the writing on the wall. He took Gilruth's advice to heart and almost immediately announced the goal of reaching the moon.

But we are getting ahead of our story. First, even for the initial Mercury flights, a new kind of test pilot was obviously needed, and in November 1958, it was decided that six should be selected and trained. They would be the best of the best. The plan, finalized in December, called for the half dozen to be chosen from an initial pool of 150 applicants, winnowed down to 36, and then trimmed to a final dozen who would actually go through the intensive nine-month training. Only then would the top half of this group be invited to the new astronaut corps.

Males alone were invited to apply. All had to be between the ages of twenty-five and forty, and stand no more than five feet, eleven inches tall. They would have be in perfect physical condition, hold at least an undergraduate college degree, and have either three years' engineering experience or equivalent flight time. Their applications for government service, levels GS-12 to GS-15, offered a salary that started at $7,000, with a top salary of $12,770.

But President Eisenhower, who was already pulling money from the manned program, scrapped those selection guidelines within six weeks and briefly put everything in limbo. Eisenhower felt that the country had plenty of military test pilots who would automatically qualify. Why pay anyone from the outside? In January 1959, NASA published new recruitment criteria. Now, in addition to the old requirements, applicants had to be test pilots and they had to have accrued at least 1,500 hours experience in jets.

Years later, Bob Gilruth decided, on reflection, that Eisenhower had been absolutely right: "It ruled out the matadors, mountain climbers, scuba divers, and race drivers," he said, "and it gave us stable guys who had already been screened for security." Obviously, NASA wanted brilliant, strong, daring, and motivated candidates who were nonetheless neither recklessly inclined nor psychologically unstable. It was a fine line, but a clear one.

The military branches quickly submitted a list of 508 men whom they deemed were qualified, and a three-member panel whittled these down to 110. This horde was supposed to arrive in Washington in three groups for further screening, but by the time the first two groups of 69 had been processed, it was already obvious that they had more than enough overqualified candidates, and the third batch never even got the chance to step off the train at the Washington station.

Teams of interviewers and examiners managed to trim this group roughly in half and now offered the lucky thirty-six men the chance to leave for New

Mexico for further testing. Four of these said, "No thanks," and the remaining thirty-two were broken into groups of six (plus a final two) for intensive evaluation at the Lovelace Clinic in Albuquerque, a logical place for medical screening since Dr. Alan M. Lovelace was widely respected as an early pioneer and advocate of manned spaceflight.

It should be remembered that at this early stage, nobody was in any way sure that humans could survive in space at all. A panel that convened at MIT, headed by a Kennedy science advisor and anti-manned-spaceflight advocate named Jerry Weisner, actually tried to have the Mercury program cancelled. The Weisner committee included a few doctors who believed that humans would die in seconds in a weightless environment. They continued to maintain that view even after monkeys had been successfully sent on suborbital excursions. Only much later, when Alan Shepard went up safely and the country went wild, did he and his panel fade to oblivion.

So first, the putative new astronauts had to be culled from the group of thirty-two, their six-person groups each spending unpleasant seven-day stretches in New Mexico. By all accounts, the men were systematically tortured. The body density of each man was gauged by submerging him in a full tank; the water that spilled out was weighed. Heart chambers were measured by having the subjects exercise with their noses blocked, and tubes down their throats appraising the air escaping from lungs. Fitness was gauged by having each man walk a treadmill that tilted one degree higher each minute, and by walking up and down a two-foot stair section every two seconds for five minutes, until his heart raced 180 beats per minute.

Applicants were measured as their feet were plunged into ice water, and as they lay on a tilted table with their heads down and the slope-angle quickly changed, and as they sat spinning in a centrifuge until their faces became distorted and blood vessels in their eyes ruptured. They were subjected to long periods of piercing noise, hours of being locked in a pitch-black room, hours of enduring a sealed chamber of 125-degree heat.

Then there were the psychological tests, with thirteen separate batteries of mental exams ("Write twenty answers to the question, 'Who am I?'").

To say that the mental screening process was thorough would be a humorous understatement. After the volunteers had been pricked, probed, roasted, and spun in nauseating centrifuges, one might assume that they'd automatically be psychologically motivated and tough, merely to agree to stick around. Nearly all later recalled that they most dreaded the hours of interviews with psychiatrists. Of course, these were military officers who already knew the ropes. None was stupid enough not to spot a trick question from a shrink. Still, when a potential astronaut grew self-confident enough, he'd find it hard to resist having a little fun. When a psychiatrist showed one candidate a completely blank sheet of white paper and then asked what he saw, the future astronaut said, in effect, "Twelve polar bears fornicating in the snow." (This was a navy man, so he didn't actually use the term "fornicating.") Given the same blank-sheet question, another told the psychiatrist, "You're holding the page upside down."

Eighteen candidates managed to make it through these ordeals with top marks. Their names were ultimately presented to Bob Gilruth, who found himself able to select seven but could see no way to winnow it down to the required six. By no medical or psychological criterion was any member of this seven better than any other—and that's how the original astronaut corps came to be seven, not six.

○

# CHAPTER 5

# The Originals

*Everyone is a moon, and has a dark side*
*which he never shows to anybody.*

—Mark Twain, *Following the Equator*

In the hierarchy of test pilots, being selected as an astronaut was the pinnacle, the equivalent to being seated at the right hand of God. Although NASA eventually put out future calls that chose dozens more, the original group forever enjoyed unique status. Even now in the twenty-first century, a half century later, their names remain familiar: Scott Carpenter, Gordon Cooper, John Glenn, Gus Grissom, Wally Schirra, Alan Shepard, and Deke Slayton. There are cities whose names are less well known than those of these men. With great fanfare, the "Mercury 7" (later and forever known as the "Original Seven") was presented to the world on April 9, 1959.

They made the cover of *Life* magazine, and were paid by that magazine for their "life stories"—only a few thousand dollars, but a lump-sum fortune to many of them, far more than their annual military salaries. They also gained the envy of test pilots everywhere. More than a few of the latter had unconcealed bitter opinions about these new idols, and regarded the Mercury capsule as requiring no more skill than a rat displays when stuffed inside a container. They called the Mercury astronauts "talking monkeys." To their minds, astronauts were passengers; they weren't really flying the craft. It took less skill to sit in a Mercury capsule than pilot even the simplest jet aircraft. This peer attitude would later do a 180-degree changeover with the Gemini program, in which the crews really did control the craft and perform extremely difficult and complex maneuvers.

Yet even in the initial sour-grapes era of the post-selection but preflight years from 1959 to 1961, the mere rigor of the screening and selection process made the small handful of successful candidates an inner elite that had no peer.

It was therefore not surprising that when a second call for candidates went out in April 1962, with "five to ten" new astronauts to be chosen, NASA was nearly swamped with applications. Requirements were again unbelievably stringent: Candidates had to be certified test pilots, but this time they could be civilian, not just military. They had to have a degree in either engineering or biological sciences. The age restrictions stayed as before, but the height requirements were slightly loosened, by one inch. Candidates merely needed to not exceed six feet tall.

To help push through their own men and ensure that they would not be underrepresented in NASA, the air force's applicants were first hurriedly sent to what the men jokingly called "charm school." Here they were coached in how to make an impression on the NASA selection people, how to dress, speak, and sit properly, even how to stand with their hands on hips (thumbs to the rear). They were prepped in how to answer questions, and even how to drink at parties (hold the half-full glass for a long, long time). Then they were hustled off to the real thing.

Candidates were sent to Texas, no longer New Mexico, for a five-day physical exam, filled with almost nonstop indignities such as having cold water poured into ears and having a metal pipe stuffed twelve inches up the anus, and endless blood tests and stamina tests and finally the dreaded battery of mental tests. Intelligence testing showed the average qualifying candidate to have an IQ of 132 on the Weschler scale, which happens to be the minimum needed for acceptance to Mensa, the "genius" society. Candidates also had their backgrounds thoroughly investigated, and were disqualified for anything much more serious than a speeding ticket.

When it was all over, nine new astronauts emerged, including names that would later be known worldwide for piloting Gemini and, especially, Apollo missions to the moon. The winners: civilians Neil Armstrong and Elliott See; air force men Jim McDivitt, Frank Borman, Tom Stafford, and Ed White; and navy men John Young, Pete Conrad, and Jim Lovell. These quickly were dubbed "The Next Nine" or simply, "The Nine." The nation now had sixteen astronauts, or fifteen if we subtract poor Deke Slayton, grounded because of an inconsequential heart murmur. (Slayton was much beloved by his peers; they even collectively petitioned John F. Kennedy to intervene when they were his guests and had his ear. But the president deferred to the doctors, who were unrelenting. Instead, Slayton was picked to be astronaut chief; from this point on he'd choose the men for each mission—not too flimsy a compensation.)

These nine picked in 1962 received much attention, but nothing that even approached the fanfare that accompanied the Original Seven. Moreover, the new total of sixteen was more individuals than could be easily personalized by the press, more than the heroes on one's local football team, and fame would now only trickle to these men over the course of years, if and when they piloted a particularly newsy space mission.

As it happened, certain Gemini missions (involving the first spacewalk, for example) received widespread publicity; others (involving later spacewalks) got very little. It was a pattern sadly repeated with Apollo. The first visit to the moon yielded a pair of names still known by everyone; the second through sixth landings were crewed by men who today are individually remembered only by science geeks and those involved with aerospace. Does the average person know the name Gene Cernan, let alone recall his appearance? How many can recite what exactly he accomplished in space? (Answer: He was the final person to walk on the moon, and one of the few who went there twice.)

Another year later, on June 5, 1963, a further call went out: NASA was going to choose yet another round of astronaut candidates. This time the requirements had changed—slightly. Now, being a jet test pilot was "preferred" rather than "required," which meant that some new people outside the insular universe of U.S. test pilots might apply. This could theoretically open the door, just a crack, for additional useful talents and specialized skills.

It was the usual grueling process, where Gilruth's NASA selection committee winnowed 271 fully qualified candidates down to 34, who went off to Texas for the customary torture. One candidate, Michael Collins, was trying for his third (and, he realized, his final) shot, since his age was approaching the upper limit. When the dust lifted, fourteen new astronauts emerged—including Collins.

The press, looking for a handle on this growing elite corps, tended to call the Original Seven the "Mercury astronauts," the next nine the "Gemini astronauts," and this new group of fourteen the "Apollo astronauts." In reality, all active astronauts were rotated and selected for whatever missions were upcoming. Gus Grissom, one of the first seven, flew Mercury and Gemini and was at a launchpad practice for Apollo when he died. And he wasn't alone in flying in all three programs. Of the first three-man crew to go to the moon, one was a member of the Nine (Neil Armstrong), while Mike Collins and Buzz Aldrin were alumni of this newest "Fourteen."

(The Fourteen had a bizarre mortality rate. Within three years, nearly one-third of them were dead of various causes, without having made a single spaceflight.)

In any case, that's how it came to be permanently known in NASA circles. If anyone said, "He's a Fourteen," people understood. Now, with a total pool of seven plus nine plus fourteen—thirty astronauts even—minus one for Slayton, NASA had enough men to get through Gemini and Apollo and go to the moon. True, a few wound up performing disappointingly poorly in space, or were grounded for other reasons such as insubordination, so that their first flight was also the last, but at least they had flown, which was more than could be said of some members of the later groups.

Subsequent years would see more candidates selected: the Original Seven, the Nine, and the Fourteen were followed by a fourth group of six who were strictly scientists; only one of these (Jack Schmidt) ever got to fly an Apollo mission. Then, another year later, the largest group of all was inducted—nineteen—of whom twelve would indeed get seats in one of the eleven three-man Apollo crews. Then came a group of eleven, all professional scientists, none of whom flew Apollo, followed by a final seven who were transferred from the Manned Orbital Laboratory after it was closed.

Seniority counted, and for the final two batches of recruits picked by the time Apollo was halfway through its run, the prospect of going to the moon had already passed them by. They could only hope to see outer space aboard the Shuttle. Still, of the seven groups of successful candidates, forty-three of these seventy-three astronauts flew in either Mercury, Gemini, Apollo, Skylab, or Apollo Soyuz. But we are now way ahead of our story.

The initial pool of twenty-nine men—those of the first three groups—supplied the majority of those names later plastered in newspaper headlines during the three incremental manned programs that put people on the moon.

Were they all cut from the same mold? In "Apollo Expeditions to the Moon," a 1975 NASA publication compiled by Langley director Edgar Cortright, Robert Sherrod gathered some interesting statistics about this select group of men, in an effort to find any commonalities that might exist. Bottom line? There were indeed striking similarities.

The men, by and large, were Midwesterners. Nearly half came from just the four states of Texas, Ohio, Illinois, and New Jersey. Half attended either the U.S. Naval Academy (eight) or West Point (five). Only one (Pete Conrad) attended an Ivy League school, Princeton.

The astronauts tended to be more formally religious than the average U.S. citizen, with twenty-three Protestants and six Catholics; many were very active in their churches. They tended to enjoy participating in (rather than watching) sports and definitely liked sports cars. Only a few spent much time with cerebral pastimes such as chess or reading classical literature. (Michael Collins was into both.)

On average, the astronaut was not in any way a youth: the average age at the time of first spaceflight was thirty-eight-and-a-half years.

Nineteen had brown hair, one black, two red, and seven were blond. Six were seriously balding.

Twice as many of the men had blue eyes as brown (sixteen versus eight) and the hazel or green category had five representatives. There were no blacks, Hispanics, or Asians in the group, and no women. None of the men were anything but Christian and none described themselves as atheist.

Three had doctoral degrees. Seven of the twenty-nine were left-handed, which is double the number that would be expected by chance.

As for the "first-born son has a higher chance for success" notion, the Apollo astronauts would seem to strongly support such a thesis, since twenty-seven of the twenty-nine were "oldest sons." In fact, twenty-two of twenty-nine were the oldest child, period, without even having an older sister. Only two had older brothers.

"Apollo Expeditions to the Moon" also includes an appraisal by psychologist Dr. Robert Voas, who was director of the Mercury training program. Despite the astronauts' almost universal loathing for psychological testing, particularly among the earliest groups in the late 1950s and early 1960s, the men would probably have had no quarrel with Voas's observations about what characteristics his staff was seeking when they were evaluating candidates. Voas recounts that he and his team were looking for "intelligence without genius, knowledge without inflexibility, a high degree of skill without overtraining, fear but not cowardice, bravery without foolhardiness, self-confidence without egotism, physical fitness without being muscle-bound, a preference for participatory over spectator sports, frankness without blabbermouthing, enjoyment of life without excess, humor without disproportion, and fast reflexes without panic in a crisis."

Lofty qualities indeed, in these men who, alone among the billions, were destined to leave the planet.

## Astronauts Who Died Early in the Program

Considering the low number of initial astronauts chosen, an extremely high number of them died within a few years. This daunting mortality rate—which mostly afflicted the two groups chosen after the Original Seven—was more in keeping with the actuarial odds of the test pilots that they were, rather than the supposedly safer astronaut corps. As of 1967, before any manned Apollo mission actually flew, eleven astronauts had perished.

Three—Grissom, Chafee, and White—died in the infamous *Apollo 1* practice launchpad fire detailed in its own chapter in this volume. Michael Adams, Charles Bassett, Theodore Freeman, Robert Lawrence, Russell Rogers, Elliott See, and Clifton Williams all died while piloting jets. Edward Givens was killed in a car crash.

This high aviation mortality rate caused NASA to alter its thinking. Prior to 1968, it encouraged astronauts to fly jets as often as possible to keep their skill levels high. After that year, it discouraged unnecessary flying.

CHAPTER 6

# Someday Alice.
# *Pow!* Straight to the Moon.

*Romeo.* Lady, by yonder blessed moon I swear
That tips with silver all these fruit-tree tops . . .
*Juliet.* O! Swear not by the moon,
    the inconstant moon,
that monthly changes in her circled orb,
Lest thy love prove likewise variable.

—William Shakespeare

In a less enlightened era, Jackie Gleason's Ralph Kramden character would threaten to send his wife, Alice, to Earth's nearest neighbor. It was in keeping with centuries of idle threats and folk songs in the "cow jumped over the moon" tradition. When quite suddenly, in the early 1960s, it looked as if some people were actually going to travel there, it didn't take too much imagination to guess who they might be.

They would be men. White men. And not just a random sample of athletes, scientists, or politicians. They'd be test pilots, which meant career military officers.

As we saw in the last chapter, then, the biographies of the first astronauts were not terribly dissimilar. The much later Shuttle era featured a lengthy period when specially selected "ordinary citizens," such as politicians and teachers, were allowed to fly. Not so the Mercury/Gemini/Apollo period. These were to be dangerous craft that required the specialized skills and raw nerve of the test pilot. And while a single astronaut—the final one on the moon—was indeed strictly a scientist, all the others were stamped from identical iron-strong "right stuff" dies, and nobody in or out of NASA had a problem with that.

Unlike the cosmonauts of the Soviet space program, no secrecy surrounded the American astronauts. They had emerged into public view, like the Space Race itself, from the rapid U.S. response to the two Soviet satellites sent aloft in 1957. When the United States announced it was televising—live—the attempted launch of America's first satellite, it was a genuinely brave thing to do. Unfortunately, TV images of this violent and dramatic *Vanguard* failure on December 6, with its disconcerting explosion followed by the rocket's nose-cone popping off on a hinge like a comical jack-in-the-box, yielded arguably the worst public relations disaster in world history. Viewers were appalled, and the world's reaction was somewhere between disbelief and laughter.

The space-incompetent Eisenhower administration then swallowed its pride and ordered Wehrner von Braun—whose team had had a rocket all ready but had been refused permission to launch—to now proceed. Sure enough, on the final day of January 1958, the *Jupiter-C* took off flawlessly and achieved Earth orbit. America was back in the space contest. The U.S. media dutifully played down the fact that America's payload was merely a spherical metal radio transmitter the size of a cantaloupe, while the Soviets had already sent up a scientific package weighing one-and-a-half tons. It was just a matter of time before the Russians would send up a person.

In response, Eisenhower created the National Aeronautics and Space Administration a few days before Sputnik's first anniversary in 1960. The idea was to place America's go-to-space effort in civilian hands, and to unite disparate councils, talented individuals, laboratories, and, eventually, the navy's space research team and the army's Alabama program led by von Braun. It was a logical idea, and a smart one.

In the last chapter we looked at the selection process that picked the men who would fly the first manned spacecraft. And how in that first year of 1958–1959 the scores of volunteers had been winnowed to the hallowed half-dozen plus one, which for decades afterward would be alluded to reverentially as the "original astronauts" or some variation thereof. With much fanfare and front-page headlines, NASA introduced the "Mercury 7" on April 9, 1959. The name Mercury, chosen for the swiftest of the classical Roman gods, was to be the program that would launch a single man into orbit in a cramped seat in a claustrophobic space aboard a modified Redstone rocket.

Although instant celebrities, several of the men were taciturn by nature, barely grunting "okay" when they were feeling particularly vocal. By and large, the Mercury Seven were not great TV personalities. By the non-slick standards of the late fifties it hardly mattered.

As it turned out, none of these men were destined to be the first in space. Two years and three days after their gala *Sputnik* introduction, the Soviets sent handsome Yuri Gagarin up, not just beyond the atmosphere but into actual orbit. Another coup from the East.

But this time the United States was not too far behind. A mere three weeks after Gagarin's launch, on May 5, 1961, thirty-seven-year-old Alan Shepard became the first U.S. astronaut to venture beyond the atmosphere, technically into space. He didn't orbit, but he did soar 115 miles up, experience the illusion of weightlessness for five minutes, and had his capsule plunge down into the Atlantic 300 miles from where he started.

That was more than good enough for ticker-tape parades and a White House visit. Seeing the cacophonous clamor, President Kennedy, less than three weeks later, made his now-famous "go to the moon before the decade is out" speech.

In today's climate when presidential prognostications are given the same general weight as those of Punxsutawney Phil, it's hard to assess how many average citizens in 1961 actually believed that Americans would indeed be walking on the lunar surface that same decade. But things were different then, and, in any case, the gauntlet had been thrown down, and the race was on.

That first successful Mercury capsule had been named *Freedom 7*, spawning endless numerological second-guessing. Why call the first manned U.S. space capsule *7*? Why not *1*? It had been Shepard's idea. His capsule had been the seventh copy built, it would be hurled aloft by the seventh completed Redstone rocket, and, possibly most important, it was flown by a member of the seven-man astronaut squad. Simple.

## Zero Gravity?

In 1961, Alan Shepard became the first American to experience "zero gravity" during a spaceflight. Later, all the Apollo astronauts felt it except when they were on the moon itself, and these days, it's a routine and continuous occurrence during extended habitations in the

International Space Station and during the week-long space adventures of the Shuttle. But did anybody truly escape Earth's gravity, or was it merely an illusion?

Gravity falls off inversely with the square of distance, but astronauts orbiting 200 miles up are not terribly far from our 8,000-mile-diameter planet. Does gravity really fade to zero at such a close proximity to the ground?

Answer: Not at all. If a skyscraper could be built that high, people standing on its roof would experience nearly the same weight that they would have on the ground floor. In truth, the only astronauts who truly escaped Earth's gravity were the twenty-seven who flew more than 80 percent of the distance to the moon, the place where our own gravity is balanced out by the moon's.

Instead, all orbiting astronauts are merely in a state of free fall. Just as a skydiver experiences "no gravity" before his or her chute opens, astronauts in orbit follow a free-falling trajectory along with their spacecraft; this imparts a feeling or sensation of weightlessness. If their craft instead could hover in place, they'd feel virtually the same gravitational pull as those on the ground.

America soon sent its second astronaut into space—the talented and taciturn Gus Grissom, who was later destined to die in the tragic and unnecessary *Apollo 1* inferno of 1967. But now, six years earlier, July 21, 1961, he tied Shepard's 115-mile altitude record and the United States was two-for-two.

An odd footnote to Grissom's trip to space still exists in the textbooks. Upon splashing down into the ocean just three miles from the intended landing spot, the Mercury capsule's hatch blew and seawater poured in. Grissom quickly pulled himself through the narrow neck and into the ocean, and then very nearly drowned as he flailed around desperately, his open space suit flooded with heavy water, while the first arriving helicopter spent its energy in a futile attempt to salvage the capsule, unaware of Grissom's plight. The module's weight vastly increased by sea-water; it was now over a half ton heavier than the chopper's maximum lifting capacity, and it sank—the first and only spacecraft lost by NASA. Subsequent investigations could uncover no reason why the explosive hatch should have detonated without Grissom's intervention, and a cloud of doubt hung over him for years. Had he panicked? Those who knew him were all of one mind: no way, not this man. Later missions justified their glowing appraisals of his character.

In any case, Shepard and Grissom had both gone up and then immediately come down. The real issue remained—how to get into orbit?

So far, astronauts had ridden atop the nine-story Redstone, which burned oxygen and alcohol in its relatively small engines. To send a heavy human payload all the way into orbit required a more powerful machine. Only one was available—a modified version of the kerosene-burning Atlas missile, which had been designed to hurl H-bombs at Russia at the start of World War III. The Atlas with its thin skin was not particularly built for safety, nor for transporting humans. Jury-rigging a few of them to blast astronauts into orbit included such temporary fixes as putting a "belly ring" around the middle, like a radial tire, to shore up its strength. Risky business.

John Glenn volunteered.

By all accounts, Glenn was the odd duck of the Mercury Seven. While most of the others lived up to the ribald reputations of navy pilots by readily availing themselves of the many Florida women who prided themselves on "sleeping with an astronaut" as if they were rock stars, Glenn would futilely scold them about reputation and faithfulness to their wives. It all fell upon deaf ears. Still, while Glenn would frequent the hangout bars favored by the other astronauts, he would never disappear into the motel with one of the always-available women.

Perhaps not coincidentally, it was Glenn who got the call to become the first American in orbit, atop the perilous Atlas.

Perilous indeed. Rockets then were nowhere as reliable as now, and explosive launch failures were almost routine. Against this nerve-wracking setting, Glenn's launch was postponed ten times for various technical reasons. It was with more than a little relief that the forty-year-old lifted off on February 20, 1962—less than a year after Gagarin. His three orbits landed him ticker-tape parades along New York's Broadway and a trip to the White House. The world was his. And he belonged to the world.

Or, rather, to the Royal Crown Cola company. Offered no further flights, he left the marines two years later and took over the helm of a company trying to gain a foothold from Pepsi and Coke. Ignominious, perhaps, but at least it was short-lived. Later he was successfully elected a senator from Ohio, then eventually ran for president on the Democratic ticket, though he abandoned the bid before the national convention.

Why didn't Glenn fly again? It wasn't due to any failing or lack of skill on his part. Unbeknownst to the media, NASA public relations experts figured that space was too risky, and Glenn too much a national hero. The cost in U.S. prestige should he perish in orbit outweighed any benefits of further flights. Let other, lesser-known champions go up instead.

Glenn's triumphant orbiting was followed by three others, with the Atlas performing perfectly each time. Everything looked great, ran smoothly.

Needless to say, that's usually when trouble begins.

# Little Things Gone Wrong

Except for the controversial sinking of Gus Grissom's capsule during the second Mercury flight, everything had gone remarkably well with both the Soviet and American manned space program. Those were the heady days of the handsome young U.S. president, and Kennedy's Camelot administration's charm seemed to have rubbed off on NASA. But lest we leave the impression that the early days of spaceflight were walks in the park for the men involved, consider the next two Mercury astronauts who were set to fly missions four and five.

First, Deke Slayton, up next to go into orbit, was suddenly yanked from the program because of a heart irregularity. His own doctor insisted that the occasional atrial fibrillation was entirely benign and would not affect his performance, but no amount of pleading could change the verdict. Slayton's wings had been clipped. Well liked by his peers, he was "elected" to be "chief astronaut," which meant, as we noted in the previous chapter, that he would do the selecting for future missions. (He finally got into space twelve years later, after Apollo had run its course, aboard the politically dictated, joint Soviet-American happy-feeling Apollo Soyuz Test Project.)

Next at the bat was Scott Carpenter. Anybody who has ever wondered whether it is possible for an extensively trained astronaut to seriously "screw up" need only review Carpenter's Mercury flight in May 1962. His performance was judged so poor that NASA officials made him the first astronaut to be immediately *not* offered another mission. For Scott Carpenter, space was a one-night stand.

According to NASA's official account, Carpenter accidentally turned on the attitude control jets not once but six times. He used even more fuel to command the craft to move in various ways just so he could take better pictures. Then he seemed to ignore the warnings from ground controllers who kept advising him that he was using up too much propellant.

Carpenter's worst errors occurred when he was preparing to return; by forgetting to switch off the manual control system, fuel was unnecessarily burned from two redundant sets of thrusters. Then he became so delayed with his checklist that the Mercury capsule pointed in the wrong direction when the retrorockets fired. As icing on the cake, he was three seconds late in firing the rockets. At 18,000 miles an hour, a few seconds does more than merely take you to the wrong side of town. The craft overshot the planned splashdown point by nearly 200 miles.

No, NASA didn't want Carpenter behind the wheel of any future spacecraft.

In the highly competitive test-pilot milieu, Wally Schirra, the very next scheduled Mercury astronaut, was obsessed with conducting his flight more than perfectly: Schirra and the controllers now wanted to see how *little* fuel could be used, and therefore how long a mission might be extended. While Carpenter had nearly exhausted the propellant in three quick orbits, Schirra landed with more than half remaining, and hit the sea just five miles from the intended point.

The sixth and final Mercury flight, in May 1963, was piloted by the last unflown Original Seven astronaut, Gordon Cooper (again remembering that Slayton had been grounded for a heart irregularity). Cooper's ride in the Atlas was to be the longest of all, thirty orbits, and everyone knew that there'd be a two-year manless gap in the U.S. program before the two-man Geminis would take wing.

This last Mercury flight was not a lucky one. On the nineteenth orbit, the first of a series of electrical problems began to plague Cooper. A few orbits later, the problems had grown intense enough so that the craft was barely functioning. It was here that the earlier astronauts' insistence on retaining some manual control proved invaluable. By receiving a complicated set of step-by-step instructions from the ground, Cooper was able to manually maneuver the capsule to a safe reentry and splashdown.

Nobody was sorry to see the Mercury program end. It was time to move on to bigger and better craft. Few suspected that the greatest challenges, most difficult work, worst in-space mishaps, and horrible loss of life, were awaiting them in the near future. Indeed, looking back now, the rickety Mercury missions were cakewalks.

○

# CHAPTER 7

# The Twins

*They dined on mince, and slices of quince,*
*Which they ate with a runcible spoon;*
*And hand in hand, on the edge of the sand,*
*They danced by the light of the moon.*
—Edward Lear, "The Owl and the Pussycat"

A quick historical glance might make the Gemini program, flown for twenty months in 1965 and 1966, seem like a letdown. After all, movie trilogies often generate a lot of excitement with their opening film and then again for their climactic final installment. People are usually less enamored of the middle feature. Similarly, Mercury first carried men into space and Apollo would take us to the moon. What was left, the filler? Who cared about the minor details of the in-between steps?

But those minor details, those "little things," happened to include absolutely everything necessary in order to go to the moon. As the Mercury program successfully if perilously concluded, no Americans had yet attempted to catch up with another object in space and then dock with it. No craft had ever changed its own orbit. Nobody had ever left their capsule for activity alone, outside, in the cruel vacuum.

And then there was the little matter of brand-new ultrapowerful rockets that had yet to be tried and tested. We weren't going anywhere until those things—among many others—had been accomplished. The concept of having two-man crews acquire the skills to attempt, and ultimately master these tasks and maneuvers, was logical and brilliant.

Alas, as with all things, if everything had gone smoothly and without incident, that preceding paragraph might serve well enough as Gemini's historical footnote. But the real fun and excitement is always when the unexpected occurs. Then and now, the unusual grabs attention, and Gemini was destined to supply it in spades.

The first attempts at rendezvous in space confounded even the flight engineers. Within Earth's atmosphere, pilots had long learned that catching up with an aircraft ahead of you is intuitive and direct: simply push on the throttle to increase power and head there in a straight line. Oddly enough, those actions produce the opposite effect when the pursuer and the pursued are free-falling in Earth orbit.

The application of thrust does indeed initially speed up your pursuing craft, but this action immediately raises you to a higher orbit so that the object you're chasing quickly falls below you. Worse, your higher orbit instantly results in a slower orbital speed, so that the thing you're pursuing gets farther and farther ahead. In short, stepping on the gas makes you slower and higher.

The actual way to rendezvous with another spacecraft is to essentially *brake*: By reducing your speed you drop to a lower orbit, which makes you then go faster so that you pass it from below. Only then do you apply thrust so that you rise up beneath it to make contact. Tricky business. And, as if on a game show, there was a nerve-wracking time limit for accomplishing all this. It was spelled U-S-S-R.

In the twenty-four months leading to Gemini, the Russians continued their series of "firsts" in space. (In retrospect, this proved to be a good thing. It kept U.S. paranoia and pride focused on the task at hand so that congressional funding never became an issue, and neither did complacency.) In mid-1963 the Soviets launched the first woman into space. Little more than year later, they sent up a team of three cosmonauts in their first Voskhod craft, and this was nearly two years before the United States was capable of launching just two people. More dishearteningly, the Russians had developed and were employing a rocket with three times the U.S. Titan rocket's power, and to top it off they accomplished a high-prestige spacewalk early in 1965, the world's first. But by then the United States was not far behind. American astronauts were able to match the spacewalking feat within two short months.

A pair of quick unmanned test flights of the new Titan (modified from the USAF's most trustworthy nuclear-tipped ICBM) were followed by Gemini's maiden flight on March 23, 1965. This was really to be a short three-orbit shakedown run with limited objectives, flown by Gus Grissom and John Young.

The technological goals were largely met; for the first time, astronauts in space were able to change the shape of their orbit. But the flight is largely remembered for a sandwich and a name.

The sandwich was corned beef. It had been smuggled on board from Wolfie's Restaurant and presented to Grissom by Young while in orbit. Fear of producing floating crumbs caused Grissom to wisely take only a few bites, but it was too late. The press coverage of the sandwich caused a furor in Congress, resulting in a clampdown of what astronauts were permitted to bring aboard future flights.

As for the "the name"—the craft had been semiofficially called *Molly Brown* after the "unsinkable" Broadway character, a deliberate attempt by Grissom to shake off the sting of the controversy that followed the sinking of his Mercury capsule, and the lingering false notion that he had panicked and blown the seals.

All previous craft had had names, most notably the "seven" series of Mercury capsules, culminating in Gordon's final, and controversial, *Faith 7* that NASA had only reluctantly allowed. (That was a different era, when church and state really were supposed to be kept separate.) NASA didn't like *Molly Brown* either, but they liked it better than Grissom's alternative choice, *Titanic*. At any rate, officials decided that after this flight the plug would be pulled on any further spacecraft naming, and the Gemini craft thereafter assumed dull letter-and-number designations. As the saying goes, bureaucracy is the epoxy that greases the wheels of progress.

The next Gemini flight provoked more media attention than any other launch; in terms of sheer publicity and excitement it was not to be equaled until the first Apollo moon-landing four years later. Because the newly in-augurated satellite service allowed Europeans to watch the launch live, and because this was to be the first launch run by the spanking-new Mission Control facility in Houston, and because the flight would include the very first U.S. spacewalk, this flight—piloted by James McDivitt and Ed White—was launched into the brightest limelight of any Gemini mission.

The June 3, 1963, liftoff was flawless, but the world was really waiting for White's spacewalk. Although he thoroughly enjoyed the twenty-one-minute EVA (extravehicular activity) and even the four minutes of propellant that his little thruster-gun provided, the actual knowledge gained was very inconclusive. White's heart had been racing the whole time, his visor had started to fog up with the exertion, his efforts at maneuvering using the tether were clumsy and unproductive. Although no one could yet be sure, this was the first real hint that while leaving a spacecraft would be safe, the act of weightless maneuvering would be exceedingly difficult.

The next flight, *Gemini 5*, got less attention but was far more important to the goal of reaching the moon. Blasted aloft on August 21, 1965, Gordon Cooper and Pete Conrad set a significant endurance record by orbiting Earth for eight days. This was longer than all previous U.S. flights combined, and longer even than any Russian cosmonauts had achieved. It really was a milestone moment, arguably the one in which the United States finally passed Russia in the Space Race. The U.S. lead would remain for the next decade, until post-Apollo American lethargy allowed to Russians to catch up once more in the field of manned spaceflight. But it wasn't easy, physically. Conrad later recalled that *Gemini* was torturously cramped: "I could hardly move. We really were in a garbage can. I'm not that big and I couldn't even stretch out full-length. It was a long eight days. You can't go out, so you just watch the clock."

But now, summer of 1965, having finally beaten the Russians in terms of space endurance, the most important goal wasn't so much competition as seeing if humans could really withstand extended periods of weightlessness. Earlier flights and Russian studies had given much reason to be concerned; heart irregularities, bone loss, nausea, and sleep problems seemed the rule and nobody knew whether there was a threshold time in space, after which astronauts would return to Earth's gravity only to promptly drop dead from some unforeseen physiological response.

It was with no small relief, then, that this crew, who did *not* exercise remotely as much as they'd been urged to, who were together the most slovenly crew ever blasted off the planet, and whose early fuel cell problems in space nearly scrapped the mission before a single day was over—that this crew successfully fulfilled the entire marathon mission. NASA's chief physician Berry was exultant: "Now men can go to the moon!"

Next came an oddity in the space program: *Gemini 7* was launched ahead of *Gemini 6*. This numerical curiosity arose because *6*, crewed by Wally Schirra and Tom Stafford, had been sitting on the launchpad set to go when they got word that the unmanned Agena craft that they were to attempt to dock with, had failed to achieve orbit. Why go to the party when your date wouldn't be there?

Instead the whole thing was delayed six weeks to December 4, when *Gemini 7*, piloted by Frank Borman and James Lovell, blasted off for what would be the new space endurance record of two weeks. It was not particularly pleasant. Both astronauts slept poorly, one had chronic constipation, and neither enjoyed remaining unbathed, unwashed, and living in the same long johns for two straight weeks in the tiny cramped capsule, unable to move. Contrary to the romantic fantasies of spaceflight, the crude toilets, constant odors, often-cold module, and lack of mobility made a prison's solitary confinement a vacation paradise by comparison.

Less than a week later, Schirra and Stafford on *Gemini 6* were ready to try again for a rendezvous, this time to meet *Gemini 7* itself.

As they sat in their couches and the countdown proceeded to zero, the two large first-stage rockets fired, and they were on their way . . . or not. Because 1.2 seconds after the engines roared to life, they abruptly shut down. Suddenly there was silence. And dread. Many expected that the next moment would offer up an enormous explosion, and Schirra grabbed and held the D-ring that would violently eject them from the top of the rocket.

Nobody had yet had to use this ejection system, and it posed a high risk of injury or even death. But an exploding rocket was certain death, and Schirra sat poised and ready to yank the chain upward toward his chest. Everyone at mission control expected him to do so. But as the seconds ticked by in silence, and the rocket hadn't yet exploded, Wally Schirra's cool hand and intrepid judgment proved right. Investigation showed that a dust cap had been left on a critical component, and, fortunately, the engines had shut down only an instant after the rocket had started lifting off the ground. It had actually risen about an inch. Any higher and it would have settled back too hard to avert disaster. The rocket was easily rerendied and just three days later, on December 15, Schirra and Stafford were finally in space. Their rendezvous with *Gemini 6* went beautifully.

The year 1965 thus ended with no deaths, no losses, and the United States well along on its goal of reaching the moon in the sixties. Sadly, this unblemished record would soon end.

○

# CHAPTER 8

# Close Call—and a Cigar

*. . . and smiling you answer "everything
turns into something else, and slips away . . .
(these leaves are Thingish with moondrool
and i'm ever so very little afraid")*

—e. e. cummings

On March 16, 1966, Neil Armstrong—a name the world did not yet know—along with David Scott, suffered the first "bring the ship down or die" emergency situation. Their *Gemini 8* flight was scheduled to last three days but actually came down after only ten hours. Interestingly, they actually accomplished their primary objective of meeting and physically docking with an unmanned Agena target.

The docking had gone perfectly. But twenty-seven minutes later, the two joined craft started tumbling end over end, faster and faster, like a berserk carnival ride. On the night side of Earth and out of communication with the ground, the crew had only themselves to help with this major, rapidly evolving, problem. Later, flight engineers confessed that they thought they would lose the crew.

Armstrong logically figured that the problem was with the Agena. He blasted his twirling craft straight back from the other, but to his horror found that the tumbling only worsened. Whatever was wrong, it was with his own Gemini module.

When over the ocean off China, Armstrong finally got in touch with the ground via a U.S. relay ship: "We're toppling end over end but we are disengaged with Agena." The tumble grew more violent, and communication with the ground was marginal. The men were growing dizzy, the ship was making a complete spin every second. Armstrong flipped all the thrusters to make sure none was inadvertently on. He radioed the ground: "We can't turn anything off!"

The astronauts' vision blurred. They were close to blacking out. If they had, with the ship spinning faster and faster, the end result would have been two lifeless bodies in space. Armstrong tried a final ploy. He disconnected all their thruster controls and deployed the reentry system. This was *only* supposed to be used when ready to return to Earth, it was breaking the rules, but he had nothing else left. And it worked. Skillfully using the new thrusters, Armstrong stopped the tumble. Now they were on their way to an unexpected emergency landing in the Pacific, but they were alive.

After the problem had been determined (the number-eight thruster had electrically stuck in the "on" position), flights were okayed to continue. But the inauspicious start to 1966 proved merely a harbinger of further trouble.

First, both astronauts scheduled to fly the next Gemini mission were killed in an airplane crash on February 28. Then, when the newly assigned replacement crew, Tom Stafford and Gene Cernan, was set, the troubles still didn't end. The main mission objective was rendezvous and docking, but the Agena craft, *Gemini 9*'s target, failed to reach orbit after its May launch and the mission was postponed for a later date. When the new target vehicle was launched on June 1, with *Gemini 9*'s launch finally arriving two days later, fortune refused to change its tune. The skilled crew managed an impressive rendezvous, but when it came time to approach the Agena for a docking attempt it was clear that something was very wrong.

"It looks like an angry alligator," one of the astronauts exclaimed, and the jawlike, partially open shroud that prevented any docking attempt did indeed seem like a demon. On to the next task, the most thorough investigation of EVAs to date. It was Cernan's job to assess the use of a back-propulsion unit, and to attempt a bevy of tasks outside the craft. But his struggles in the weightless environment were so much harder than anticipated that perspiration first fogged up his faceplate, blinding him, then later beads of sweat simply dripped stingingly into his eyes as he gasped with the efforts of moving along and stopping where he wished to. As his legs flew over his head unintentionally, and his heart rate varied from a wild 140 to 160 beats per minute, he generated more moisture than his suit's air conditioner could remove. Cernan suffered repeated problems with his faceplate fogging until finally the EVA was called off short when both astronauts realized that it was becoming dangerous and largely futile, and NASA was getting yet another lesson that activities outside a spacecraft were far more taxing and difficult than anyone had ever imagined.

*Gemini 10* was a far more satisfying affair, a turnaround toward better times. Piloted by Michael Collins (who would later be chosen as part of the first three-man team to the moon on *Apollo 11*) and space veteran John Young, it used *Gemini's* ace card (ability to maneuver by the pilots rather than from ground control) to full advantage.

First the target Agena was launched, followed by their *Gemini 10* craft, which flawlessly caught up with it and docked. Then, after linking together the two components, they ignited the powerful Agena engine and experienced a forceful, frightening propulsion that lifted them to a new altitude record of 458 miles.

## Eyes In or Out?

All of NASA's first two teams of astronauts had been test pilots, and were accustomed to dramatic thrusts of energy from powerful engines. Such powerful g-forces could push them either back into their seats, which they would describe as an "eyes in" force because they could feel their eyeballs uncomfortably pressed into their skulls, or "eyes out," which was even worse because it made them feel as if their eyeballs wanted to pop from their heads. Eyes out happened during sudden or dramatic braking, during certain aerobatic maneuvers, and were also specifically mimicked by the torture machines in which astronauts were trained. During actual missions in space, almost all acceleration was of the eyes-in type, with one notable exception. During the *Gemini 10* flight, the two pilots attached their craft to the powerful Agena rocket in a manner that had them facing the great engine, giving them a backward ride. It was the only way to link the two, in order to hurl Gemini into a much higher orbit. When fired, this rocket produced such an unexpectedly violent and dramatic eyes-out acceleration that astronaut Mike Collins later confessed that, "I almost shut it down."

The experience, even for these two veterans, was disconcerting. Later, in his book *Carrying the Fire*, Collins describes the ignition of that Agena rocket engine: "Aw shucks, I think, it's not going to light, when suddenly . . . I am plastered against my shoulder straps. There is no subtlety to this engine, no gentleness in its approach." He later said that the acceleration felt much greater than it was supposed to, and that it was "very uncomfortable."

In fact, after the burn, as the craft hurtled upward toward its new altitude record, Collins turned to Young:

"I almost shut it down."

"No, you didn't."

"I almost did. If you had said shit, I would have shut it down. Really."

After most of a day at the record height, they burned the Agena again and returned to a more usual 180-by-240-mile elliptical orbit, where Collins opened the hatch to perform the world's first ultraviolet measurements of hot stars. In later years, special UV telescopes would take over the job; however, since most ultraviolet wavelengths cannot penetrate Earth's atmosphere, and can only be monitored from space, *Gemini 10* provided the very first peek at these sizzling blue suns.

Before they could get started, however, the crew got a reprimand from Mission Control. They had been too quiet. The mere promise of later scientific papers would not do; the press (and NASA's public relations department) was hungry for immediate news headlines; they wanted these no-nonsense guys to stop doing their chores in silence.

"Okay, what do want us to talk about?"

"Anything that seems appropriate," said Mission Control.

Such distractions aside, the crew soon emptied all the air from the cabin, opened the hatch, and Collins proceeded to float in the opening. It was night and except for occasional lightning in the clouds far below, everything was pitch black as the craft glided enchantingly beneath the stars. Collins was enraptured by the wide scope of the heavens from his perch, with his upper body outside the craft. He marveled at how the stars were perfectly steady, since the twinkling seen from Earth is caused by an atmosphere that space of course does not possess.

It was then and there that a truly frightening event began: both astronauts were temporarily blinded. It started with a stinging in Collins's eyes, outside the ship, and progressed to incapacitating tearing and blurring. Calling inside, he found that Young was similarly afflicted. Here they were, their cabin's air evacuated, two men visually incapacitated. Was it the result of a wipe-on chemical that had been applied to the inside of their visors, as was the case with an earlier fogging incident involving Gene Cernan? No—later it was found that some lithium hydroxide drying-agent was being pumped into the space suits' oxygen supply. Collins came in, they groped around for the hatch lock, managed to find the right valves to pressurize the cabin, and things slowly started to improve.

After this, *Gemini 10* went on to achieve a second successful docking with yet another target vehicle. After, Collins performed a major EVA, going out to retrieve a micrometeor shield from this second Agena spacecraft. It would help experts assess the density of small meteoroids in near-Earth vicinity, a vital piece of information for all future space walks. Although Collins was success-ful at retrieving the experiment, he found, like those before him, that maneu-vering outside the craft in zero-g conditions was frustratingly filled with constant tumbles, overshoots, lost holds, and snags in lines. In Collins' post-flight report, he particularly urged that future craft should be equipped with external handholds if astronauts would be expected to venture outside.

And of course, they *would* be expected to go out. The whole idea was to prepare for the Apollo trips to the moon, in which spacewalks would be life-saving in the event that a successful docking was not accomplished after the lunar lander returned to orbit following its time on the surface. Spacewalking would then be the only way that astronauts could get from the LM to the Command Module that would bring them home.

So important was this aspect of spaceflight that the next *Gemini*, number *11*, carrying Charles Conrad and Richard Gordon, made this its main priority. The week-long mission that began during a two-second launch window on Septem-ber 12, 1966, brought home an impressive series of "firsts." The pair performed the first docking and rendezvous (with their Agena target rocket) in initial orbit just ninety minutes after launch—simulating maneuvers the later Apollos would have to accomplish. Now, however, they did it repeatedly. They also got boast-ing rights to the first pair of space vehicles joined by a tether. By getting the two craft to revolve around a common point in space, they created the first-ever sense of artificial gravity. It was just enough—about a thousandth Earth's grav-ity—so that a loose object slowly moved back against the real bulkhead.

When attached to the Agena, the new booster threw them up to the highest-ever orbit, a whopping 853 miles, a new altitude record—more than three times the height of today's space station. The only sour note was the re-currence of an old Gemini problem: on the second day, when Gordon performed an EVA, excessive exertion made his heart race and his breathing rate speed up.

Worse, his visor fogged up and, like astronauts before him, he could no longer see. He had no choice but to come in, ending the EVA in half the scheduled time.

The EVA blind-and-gasping astronaut problem clearly had to be addressed, and there was just a single Gemini flight left. The job fell to James Lovell, along with the man many regarded as the most knowledgeable of all in the field of space rendezvous, and the only doctorate holder among the earliest astronauts—Buzz Aldrin. He would later of course gain fame as a member of the first team to land on the moon. For *Gemini 12*, however, Aldrin was determined to prove that a properly trained astronaut could do what none before had accomplished: perform spacewalks without excruciating difficulty.

To do this, Aldrin pioneered the technique of spending not days or weeks, but months of training in underwater tanks to simulate the zero-gravity environment. A perfectionist, he did indeed nail it. Exquisitely.

When he and Lovell walked to the launchpad, both men had signs attached to their backs: one said "The," the other, "End." Theirs was the last of the Gemini missions.

And indeed it ended splendidly, unlike the trouble-plagued final Mercury flight. Although their onboard radar turned balky soon after launch, Aldrin's "Doctor Rendezvous" moniker proved well deserved: he plotted the needed rocket firings and trajectories for a docking manually, without even the aid of any computer.

When it came time for the EVA, Aldrin performed a series of dozens of tasks. He torqued bolts, plugged and unplugged cables, installed a camera, and generally did the sorts of things that, unknown to him, would be among countless tasks the later Shuttle astronauts would be called upon to execute when servicing the Hubble Space Telescope and fixing other problems. Through it all—a record five-and-a-half hours outside the spacecraft—Aldrin barely broke a sweat. His heart rate remained normal, his visor did not fog, he demonstrated that humans could indeed perform well outside their craft.

Splashing down four days later on November 15, 1966, Gemini was now officially over and Americans would not venture again beyond our atmosphere for another two years. The press, the growing pool of astronauts-in-training, the public, the world—all now waited for Apollo, and its soon-to-be-completed heart-stopping rocket upon which the United States had placed its entire moon-dreams.

○

# CHAPTER 9

# Apollo's Logic

*Chanting faint hymns to the cold fruitless moon.*

William Shakespeare, *A Midsummer Night's Dream*

When Aldrin and Lovell stood on the aircraft carrier USS *Wasp*'s deck on November 15, 1966, the Gemini program was over—and the Russians had been beaten.

In the two short Gemini years, the Soviets had gone from perceived shoo-ins for winning the Space Race, to a new unaccustomed role of playing catch-up. American astronauts had now logged nearly 2,000 man-hours in space, compared to the Russians' 500. And while the Soviet Union had a nasty and worrisome habit of springing surprises, the United States was gaining confidence that maybe, just maybe, they could really pull off the Kennedy goal of standing on the moon before the sixties were over—and before anyone else did.

Everything now depended on the final leg of the triathlon, whose name was Apollo.

Of course, the Apollo program did not merely spring up like a phoenix from the ashes of Gemini. Much of the serious early planning and even contract-awarding had been done way back in 1961, while Mercury was still flying. By 1962, the entire series of planned Apollo flights had been in place. Moreover, a set of new rockets was already in the design and early-construction stage. They were ready to test, and then, if all went well, the finished product could be hurled to the moon by 1969. In order of size, and with no regard for name consistency, they were designated Little Joe, Saturn 1B, and Saturn V, and they would in turn test first a Command Module mockup, then a fully functional but not moon-capable module called Block I, and then finally the ultimate, the Block II Command Module.

It was at this time, too, that the entire elaborate get-there strategy was set for all time. The three-stage Saturn V rocket would launch a payload that consisted of three distinct, detachable segments.

The most important was the three-crew-member Command Module with its 600 switches, 64 dials and indicators, and 71 lights. It could keep three people alive for ten days, supplying them with cold and hot water and oxygen. It would remove carbon dioxide from its own air, create its own electricity, and was airtight and watertight. It could turn itself in any direction. It was the most complex piece of machinery ever created. For 99.8 percent of the mission time it would remain attached to its Service Module, a sort of in-the-basement support structure containing the large all-important refireable engine, fuel, and arrays of other vital equipment that nonetheless didn't need to be accessible during the mission.

The final component of the Saturn payload would be the insect-looking, legs-folded Lunar Module, or LM, the only part that would actually make contact with moon dirt.

As for how these three modules would be used, the sequence would go like this:

After launch and following a two- or three-orbit busy period spent checking that everything had reached space in workable condition, the crew would blast out of Earth's vicinity using the Saturn's single third-stage J-2 engine. The Command Module attached to the Service Module (the two together were called the CSM), fires the Service Module engine to remove itself from the third stage, does a 180-degree pivot in space, slowly dives into a compartment on that third stage, docks nose-to-nose with the LM, and carefully fires again to yank it backward, free from its storage area. Got all that? It's arguably the most complicated part.

After this, the CSM-mated-with-LM spends three leisurely days in free fall to the moon, firing only on rare occasions to make small course corrections to the desired trajectory. At the moon, they perform a major engine burn to slow down enough to be captured into moon orbit. A second burn soon afterwards circularizes the orbit. Then two crew members crawl through the narrow tunnel into the LM, detach from the CSM, fire the LM's single engine for its slowdown and descent to the surface.

After the moonwalk and science activity on the surface, which can last from one to three days, the upper-ascent stage of the LM fires, leaving behind its descent-stage engine and platform. With split-second timing, it rises to meet the still-orbiting CSM, docks with it, and the astronauts crawl back through the narrow tunnel.

They then eject the LM and send it on its way to crash onto the moon's surface, and fire their SPS engine to blast out of lunar orbit for the three-day coast back to Earth. Nearing Earth at a very narrow entry trajectory so that they neither travel too steeply, which would make them heat up and disintegrate, nor too shallow an angle, which would have them skip off the atmosphere like a stone off a lake surface, they ignite explosives to separate from the Service Module for the first time, point their curved heat-shield section downward, and scream into the atmosphere at a blistering 24,700 miles per hour to a splashdown in the sea.

The entire series may seem complex, and it is, but by discarding un-needed modules, engines, and other components when they are no longer needed, the journey becomes weight-logical. Of course, everything has to work. And there are one million separate parts.

# CHAPTER 10

# Personal Risk

*Nothing that is can pause or stay;*
*The moon will wax, the moon will wane,*
*The mist and cloud will turn to rain,*
*The rain to mist and cloud again,*
*Tomorrow be today.*

—Henry Wadsworth Longfellow, "Keramos"

Volunteering to become an astronaut—to fly aboard missiles that could easily explode—meant that you were either stupid or brave.

None of the astronauts were stupid.

From the get-go, even the Mercury Seven knew that they would be placed within experimental machines subject to the same faults and problems as all other complex groundbreaking equipment. Estimates of fatal-failure probabilities—usually from semiofficial sources—seemed generally too optimistic, but nobody really knew because the technology was still evolving, with problems tackled as they arose.

Aerospace engineers have always been aware that additional safety can be designed into any flying machine—at a cost of time, money, and weight. But in the sixties, the upcoming missions to the moon simply did not have much time; they were supposed to succeed in a mere eight or nine years from start to finish. As for budget, it was large but not infinite. To accomplish the goal, then, how much risk was acceptable?

Decisions about risk were common in the initial Mercury, Gemini, and Apollo design years of 1959 to 1963. The trips into space would, it was thought, be flown exclusively by test pilots accustomed to their high-peril occupation. When interviewed, these astronauts-in-training were generally reluctant even to discuss danger, stoically shrugging off personal peril. Knowledgeable science writers of the time guessed that the test pilot's traditional actuarial odds wouldn't change too much, and that the space program would be lucky if deaths could be limited to no more than one astronaut out of ten.

Was it therefore sufficient if engineers improved the odds so that nineteen out of twenty retired alive? Or should the considerable effort be invested to bring it to ninety-nine out of a hundred?

When engineers asked Bob Gilruth, director of the Manned Space Craft Center, and Walt Williams, the Project Mercury director, what kind of security and reliability should be engineered into the program, they got two very different answers.

Williams announced that he wanted to make the risk of astronaut death one in a million. Engineers present at that meeting kept straight faces out of politeness, but knew that such a level of redundancy and security simply could not be accomplished. Even airline passengers of the day did not enjoy that degree of safety. Merely to attempt such risk-aversion would delay the moon program by years, even if was technologically possible, which was highly doubtful.

"How about three nines?" Gilruth suggested. And this was accepted: the success odds were to be 999 out of 1,000. Spacecraft systems, subsystems, and rockets would be designed so that each astronaut faced only a one-in-a-thousand chance of not making it back alive.

That was the official engineering criterion. The "inside" truth, however, is that few at NASA really believed that this degree of safety had ever actually been achieved. (And unfortunately, in the fullness of time, they were proved right.) Nearly everyone expected an eventual fatal accident. There simply was too much that could go wrong.

It wasn't merely Murphy's Law. The Saturn rocket alone contained nearly a million parts, along with tons of flammable kerosene and explosive hydrogen. The Command Module had another half million parts. Already the space program had seen its share of disconcerting mishaps, as when a cover had been inadvertently left on a piece of equipment, causing the 1965 *Gemini 6* launch to shut down after the engines had already started to fire. How could absolutely nobody, of the 420,000 people working on the program, not screw up in some significant way? Was it even possible?

Very early on, engineers and planners identified which systems or subsystems simply could not be allowed to fail. For example, the single engine of the Service Module would have to be fired again and again to reach the moon, but more important it would *have* to function for the astronauts to leave lunar orbit. Unless they could increase their speed by at least 2,000 miles per hour at that time, they simply could not come home.

The single engine of the LM ascent stage would also *have* to fire or else the men would remain to die of suffocation on the lunar surface. The parachute system would also *have* to deploy, or else the astronauts would receive fatal g-forces when they hit the ocean at 350 miles an hour.

The engineering philosophy employed to ensure that critical systems would always work was to use the simplest possible technology and then build in redundancy wherever possible. For example, the truly critical engines would use a hypergolic fuel mixture of hydrazine and nitrogen tetraoxide. When these two liquids come in contact with each other they ignite spontaneously. No ignition system is required, no spark, no electricity.

The next safety step is to store the two fuel components in tanks with nylon bladders. Helium gas, released from a special tank into the space between the fuel tank and its bladder, squeezes the fuel and drives it toward the combustion chamber. No fuel pump is required, just the release of inert gas. Ball-valves also driven by gas pressure open and close to release the two fuels. Duplicate valves can take over if any fail. Simple. And it worked every time. In fact, it was hard to even imagine how these engines *could* fail.

But NASA insiders, and the astronauts themselves, also knew of Apollo's weak spots. One was the Saturn's second-stage engines, the brand new J-2s, which used the highly efficient but relatively untried mixture of explosive hydrogen and liquid oxygen. Another worrisome area was the heat shield. Mercury and Gemini missions had merely returned from Earth orbit at speeds through the atmosphere of 16,800 miles per hour. Apollo would be screaming in at 24,700 miles per hour, with air friction raising its surface to a glowing 5,000 degrees Fahrenheit. Would its 300,000 honeycombed compartments hold, and truly keep the occupants from burning up? Three unmanned tests were successful, but still. . . .

There were other weak links. One in particular that gave ongoing cause for concern was the oxygen system. Earlier astronauts had spent all their time in orbit in a low-pressure (one-third Earth atmosphere) environment of 100-percent oxygen. There were many benefits of this one-gas system, and additional advantages to maintaining an environment of low pressure, where leaks to the vacuum of space would be less likely. It had worked for six years through all the Mercury and Gemini flights. But there were worrisome signs that a pure-$O_2$ environment remained risky. Some NASA engineers kept cross-ing their fingers that the single-gas advantages would prove to be worth that risk, and they remained concerned.

The worriers included Gus Grissom, the talented, quiet veteran of Mer-cury and Gemini missions. His apprehension was not the oxygen itself, but the shoddy and rushed work being done by the Command Module contrac-tor, North American Aviation. During 1966, it became increasingly clear that this contractor was the Apollo program's Achilles heel. The relatively inex-perienced aerospace company was having difficulty delivering completed CMs and displayed problems with supervision and quality control, and this in turn was slowing down the entire launch schedule.

In a way, any onlooker could sympathize with the company and its en-gineers: the Command Module was not only the most complex machine in the entire program, and the most sophisticated ever built, but it also kept being modified and tweaked in dozens of various directions. All along, as mentioned earlier, the company's contract called for two versions. First they'd deliver a series of Block I modules that would be used only as shakedowns on the first Saturn IB launches. Then, based on results, tests, final requirements, and the improved lift capability of the Saturn V, they'd produce the finished Block II Command Modules that would actually go to the moon.

The first flights in Earth orbit required Block I models by 1966, so they were frantically being rushed to completion for their all-important mating with the other spacecraft components, for flight testing. Technicians, engi-neers, and electricians, faced with modifying instruments, retrofitting new gauges, or fixing or replacing various pumps, switches, and other compo-nents, had to repeatedly undo nearly finished CM electrical panels and splice together myriad new connections and add new lengths of cable and wire. The cramped CM, small enough to fit in a large living room, already contained sixteen miles of electrical wires. Adding endless new segments with connectors and splices produced Gordian knots of countless multi-colored snakes, and it was totally filling the unseen spaces behind almost every panel. Indeed, workers often had to use all their strength to push the free wires and wire bundles back in. Whenever any panel was unscrewed it would pop out toward the worker from the sheer outward pressure of all the packed and tangled wires.

The astronauts, who would visit North American's factory or view the later stages of work at Cape Kennedy, were often taken aback by what they saw. Gus Grissom hung a large lemon from one incomplete Command Module at the Downey, California, plant. The crew was well aware that in a pure-oxygen atmosphere, protection from fire was based solely on preventing any sparks or ignition sources. Had this really been accomplished?

In March 1966, the pilots of the first Apollo mission were chosen and announced. It was to be commanded by veteran Gus Grissom, Ed White (who had taken America's first walk in space), and the sunny-spirited space newcomer Roger Chafee. They would use Command Module number 12 to fly test mission *Apollo-Saturn 204* (later called *Apollo 1*) in its very first manned orbital trial. A month before that inaugural flight a ground test would be performed; this would include a "plugs-out" launch simulation where a fully manned and ready-to-go Saturn would have all its electrical connections with the control center and gantry pulled out. The entire spacecraft, rocket, and subsystems would be checked while under its own internal power alone.

This test was slated to take place in the fall of 1966. But when spacecraft number 12 arrived at the Kennedy Space Center in September, a large number of faults, malfunctions, and problems were discovered. When these were finally fixed, a serious design flaw with the environmental system's oxygen regulator was found, and the entire unit had to be yanked. Then a cooling leak developed.

Just as disheartening, the contractor had made 623 changes and tweaks to this spacecraft since its arrival at the Cape, and astronauts could only wonder what they were getting themselves into, literally. It must have been frustrating all around, considering that this "Block I" craft was only an interim model and would soon be replaced by flights solely using the Block IIs. Still, it had to work for the program to proceed.

After many delays, the plugs-out test was finally scheduled for January 1967. It is easy with the wisdom of hindsight to see just how dangerous this was.

After all, there had been previous accidents in pure-oxygen environments, and everyone was aware of them. On September 9, 1962, on the penultimate day of a two-week experimental session in an all-oxygen chamber, a smoky spark-induced fire erupted within some electrical measuring gauges and the two subjects had to be rushed to the hospital. Ten weeks later, on November 17, a much worse mishap critically burned two test subjects and seriously burned two others. In that case, on the sixteenth day of a three-week stay in a sealed U.S. Navy oxygen test chamber, a persistent flame erupted from an ordinary 24-volt light socket when one of the four subjects tried to change the bulb. He then asked through a microphone what he should do and was instructed to smother the flame with a towel.

Bad advice. The towel immediately leaped into brilliant and violent flame, which almost as quickly spread to and engulfed the man's clothing. Those who tried to help found themselves on fire as well. For the hundredth time, researchers were graphically reminded that slightly combustible materials such as cotton become explosively flammable in a 100-percent oxygen environment while other materials that normally do not burn at all, such as steel wool, combust vigorously. The fact was, materials placed for any length of time in a pure-oxygen environment absorb some of the oxygen and change their burn properties. Worse, a flame once started will burn five times hotter, brighter, and more vigorously than in a normal earthly environment.

It was against this perilous background that engineers' worst fears became reality on Friday, January 27, 1967.

# CHAPTER 11

# The Horrible Deaths of Mission AS-204

*. . . the vast slow-breathing unconscious cosmos with its dread abysses and unknown tides.*

—William James, *Essays*

By the end of 1966 the Apollo program seemed on target for reaching the moon in 1968 or 1969. Three previous unmanned flights aboard the modest Saturn IB, designated AS-201 and AS-202, had shown that the new heat shield design worked and the rocket and module would stand up to the stresses of flight. A third flight successfully tested the critical third stage, which would have to shut down and be restarted during actual moon missions. Now it was time for AS-204, the first manned shakedown flight in orbit, tentatively scheduled to last fourteen days. A month before they could actually get spaceborne, however, there was the little matter of a full launchpad test of all systems prior to flight.

Gus Grissom, Ed White, and Roger Chafee entered their module a little after 1:00 p.m. and Grissom quickly noticed an odd odor. The test was halted while gas samples were taken and then analyzed. It was nothing serious, and the test resumed, with the hatch closed at 2:40.

Actually, the Block I capsules had three separate hatches. First there was an inner pressure seal that could only be closed from the inside, using six bolts and a special ratchet tool. Then came the thermal shield, installed from the outside, a square section of heat shield material that could later be moved from either the inside or the outside. The final piece was merely a fiberglass cover designed to protect the windows in the event the escape rocket had to fire. It would be yanked away when the first stage finished its burn some three minutes after liftoff, and could easily swing away in the event that the astronauts had to exit the craft. In all, the astronauts could remove the three hatches and be out of the command module in sixty to ninety seconds in case of an emergency.

The six-bolt idea was far from a quick-egress system, but had been added to the module design after Gus Grissom's hatch prematurely and explosively blew in the sea, sinking his Mercury capsule. It is extremely ironic that the very same astronaut who experienced the hazard of an overly quick-opening hatch system, should now pay dearly for the result of that event—a design that had gone too far in the other direction.

The astronauts were strapped in, the hatch sealed. To duplicate all the procedures of a real launch, the interior was then filled with pure oxygen raised to a 15-percent higher pressure than that of the world outside. This overpressure was designed to prevent any outside air from entering; in the event of a leak, gases would go from inside out rather than the other way around. (Nitrogen had to be kept away from the astronauts for hours before launch; once in space in their later low-pressure Command Module environment, any nitrogen in their bodies would bubble out into their blood and joints, producing the dangerous and excruciatingly painful condition divers call "the bends.")

Oxygen quickly reached a pressure of thirty-four inches, compared with the normal average sea-level pressure of about thirty inches. Countdown tests and procedures began, but were halted repeatedly due to glitches, and then halted at greater length after three hours. The most serious and frustrating of the recurring problems was that conversations between the crew and controllers were hard to hear or even completely inaudible. Static, crackling, cutoffs, and other audio issues plagued the test. Grissom, enduring this for hours, finally uttered a complaint, and when even *that* could not be heard he repeated in exasperation, "I said, Jesus Christ, if we can't communicate over three miles, how the hell are we going to communicate when we're on the moon?"

Now it had been five hours since the astronauts had been strapped in. The clock read 6:30; it was getting dark. AS-204 was on yet another hold for more equipment problems, but the hold was scheduled to end in another minute.

Suddenly, at precisely 54.85 seconds after 6:30 p.m., telemetry showed a surge of power in electrical bus two. It was only a tiny blip compared to the total current the spacecraft was using, but it indicated a spark, a short circuit. Moreover, at that moment, all other equipment connected to this alternating current bus displayed an interruption in their functioning.

Nine seconds later, at 4.7 seconds after 6:31, the normally stoic and taciturn Gus Grissom yelled, "Fire! We've got a fire in the cockpit!"

All three astronauts surely knew at that instant that they didn't have a chance. But Ed White unstrapped himself and leaped to the hatch with the ratchet. The impossibly cumbersome design meant that he had to fully unscrew six separate bolts. He had barely begun when the fire became supernally brilliant, leaping across the module with increasing fury. With the oxygen at overpressure, dozens of objects started to burn simultaneously. Velcro—a new material that was everywhere in the cabin—ignited along with plastic netting; it all fell in glowing globs of dripping liquid flame. Outside, twenty-seven technicians on the gantry and in the white room saw the windows brightly alight, and rushed to help, only to be repelled by the heat.

Two seconds after Grissom's emergency call, monitors showed the temperature and pressure rising dramatically. The astronauts' fate was now sealed: even if White had managed to undo the bolts, the internal pressure would keep the hatch pushed immovably against the wall. The inward-opening design doomed any slight remaining hope of escape.

White sheets of fire leaped from one side of the Command Module to the other. It was seventeen seconds after 6:31, and it was Roger Chafee who now sent a desperate futile call to the controllers three miles away:

"We've got a bad fire—let's get out!"

Before he even finished speaking, the internal pressure from the inferno had reached between two and three times the normal external atmospheric pressure, and this now exceeded the design limits for the module's structural integrity. The hull split at the bottom right, giving the white-hot gases and flame a path for increased motion; the fire leaped at high speed across all of the astronauts. At this moment Roger Chafee transmitted his final cry:

"We're burning up!"

Workers outside heard a horrible roaring as hot gas suddenly gushed out of the crack in the capsule, with smoke billowing into the gantry and the adjoining white room. From within, over the radio channel, came a final cry of agonized pain from Chafee, and then silence.

A mere twenty-five seconds after Grissom's initial call, the heat had already melted the astronauts space suits around their bodies. Now, the final flames were dying out from lack of oxygen, and a choking black lethal atmosphere was all that remained inside. An astronaut still breathing would be killed within four minutes from asphyxiation, which indeed was later listed as the official cause of death.

At the height of the inferno, and when the sounds of the rupturing spacecraft could be heard, most workers thought that the entire rocket was blowing up, and nearly everyone fled. Now, minutes later, when nothing further happened, many rushed back and desperately tried to open the capsule despite several small fires that had started all around them. To the accompaniment of fierce heat exiting the inky smoke-filled opening, the hatch was finally opened five minutes and twenty seconds after the fire began.

All was black and charred inside. Using flashlights, technicians were puzzled at first, for there appeared to be no sign of the three crewmen.

It took some time to recognize and identify the horrible scene; only Chafee was still lying on his couch; the other two were on the floor. Grissom had tried to seek refuge under the middle couch, while White, after abandoning the futile effort of unbolting the door, had squirmed flat between the hatch sill and couch headrests.

There had been no fire personnel or equipment standing by during this test, and the first did not arrive until three minutes after the hatch had been opened. Doctors did not get there for yet another three minutes, and then quickly reported back to the center that all three astronauts were dead. No oxygen equipment was at hand, so nobody could remain in the hot interior for more than a few seconds, but later when the air had cooled, technicians made a grisly discovery. The bodies simply could not be freed. The men's spacesuits had melted and fused with melted nylon at the height of the fire; all this combined plastic had resolidified to form a hard pool, like frozen lava, welded solidly to the floor.

Gus Grissom's suit had fared the worst, with three-quarters of it melted off. The effort at freeing the bodies, which began soon after midnight, some six hours after the fire, took ninety minutes.

By the next day, other Block I modules were impounded and an inquiry begun. But beyond the deep tragedy to the men's families, and the nation's sadness, came the tacit question in everyone's mind: Was this the end? Could we still reach the moon?

○

# CHAPTER 12

# The Apollos Finally Fly

*We will sing of the vibrant nightly fervor of arsenals and shipyards blazing with violent electric moons.*

—Tommaso Marinetti, "The Founding and
Manifesto of Futurism"

Like a wet dog shaking itself dry, the United States refused to stand still after the horrific launchpad fire. Not even a single day was permitted to pass, before calls to action were sounded.

True, the days following January 27, 1967, found NASA in a deep funk. But the inevitable investigations and congressional hearings were already in motion, and everyone knew that this would be no whitewash. The largely science-illiterate public didn't know it, of course, but the Rube Goldberg Block I command module was about to be replaced by the vastly improved Block II model anyway. All that remained was to incorporate the lessons learned from the fire into the new version.

Still, those who followed the investigation had to be taken aback by the previous lack of supervision, even downright negligence by the module's builder, North American Aviation, and NASA's lack of effective oversight. Gus Grissom had been right to be concerned: now the miles of jury-rigged wiring, impossibly cumbersome escape hatch, ultraflammable high-pressure oxygen environment, and lackadaisical attitude toward astronaut safety was sobering and disquieting.

But Gus Grissom's own words were also publicized and taken to heart. The fallen hero had said, "We are in a risky business and we hope that if anything happens to us, it will not delay the program. The conquest of space is worth the risk of life."

Well said: everyone needed to roll up their sleeves and move on. By the time it was all over, the Block II capsule underwent 1,300 changes. No longer would astronauts on the ground be in a 100-percent oxygen environment. Every bit of material would be fireproof. The hatch was completely redesigned, and now could be opened from the inside in just five seconds. And it would open outward, not inward—a basic safety lesson that had already been incorporated into most municipal codes for stores and other public buildings. (Investigators also revealed that even without the fire, the inward-opening *Apollo 1* hatch would have been unopenable for the simple reason that any interior pressure greater than 17 gm/cm 3 above the outside Earth-atmosphere pressure would hold it firmly pressed closed against its seals. During the test, the interior pressure had been raised to 140 gm/cm 3 above ambient pressure even before the fire vastly increased this further. In short, unless they first took the time to vent the interior gas, the doomed astronauts could not have opened their hatch no matter what they did, period.)

In fairness, it should be added that the investigating bodies found the CM's deficiencies, and the fire's ultimate cause, to be ignorance and lack of awareness rather than any malevolent cost-cutting or criminal substitution of inferior materials.

Among the changes, the potentially flammable glycol (cooling) liquids would now run in pipes that were vibration-tested against leaking solder joints. Conduits that were aluminum (which can burn in a pure-oxygen environment) were replaced with incombustible stainless steel. Some 2,500 potentially combustible sources in the module were replaced with fire-resistant analogs.

Given their burned and melted states after the fire, the space suits were redesigned as well. The Nomex material that composed the suits worn by the expired Apollo crew was tested, and found to burn at 900 degrees Fahrenheit. It was replaced with new suits made of Beta cloth, which would never propagate flame at all, and wouldn't even begin to melt until it reached 1,550 degrees Fahrenheit. Anyone wearing it would even find themselves able to safely resist direct flames for several seconds.

A long debate began about what should be done with the pure-oxygen environment. Even with every possible ignition source eliminated, and completely fire-resistant materials in the module, an overpressure-oxygen environment on the launchpad was now eliminated and replaced with a sea-level-pressure mixture of 40 percent nitrogen and 60 percent oxygen. To keep nitrogen from the astronaut's bodies, occupants awaiting launch would now remain in their suits, breathing pure oxygen until and beyond liftoff.

As for the mission in space itself, this was far more difficult to change. Myriad systems had been designed around the one-gas system, so this was permitted to remain once the astronauts were aloft.

A pure-oxygen environment in space might at first seem an even worse situation than one on the ground, but it was actually quite different, and much safer.

First, the in-space module environment would only be pressurized to one-third the sea-level pressure of Earth. Such oxygen molecules are four times more separated from each other than they were in the *Apollo 1* launchpad atmosphere, and fire is far more reluctant to spread.

Second, in space, astronauts need only don their suits and helmets and vent all the gas from the capsule to space. The fire would be instantly extinguished in the resulting vacuum.

Third, fire acts very differently in a weightless environment. On Earth hot air rises, which is why hot air balloons ascend, why candle flames have an upward tapering appearance, why smoke rises, and why campfire flames leap up toward the sky. It's because hot air is less dense than its surroundings.

Rising up speedily, an earthly flame carries away its own products of combustion such as carbon dioxide, and the upward motion also creates a partial vacuum near the flame that pulls in new oxygen from the sides, which aids continuing combustion. It's a self-propagating affair. None of this happens in zero g's. There is no "up," so a flame assumes a spherical appearance. The carbon dioxide is not carried away, and therefore lingers to smother the fire. No new oxygen is drawn in to keep the fire going. In weightlessness, fires tend to quickly go out all by themselves. (Note: A slow, lingering smoky situation is another story. In later years, Russian craft have experienced two separate hazardous smoke-filled environments that had to be dealt with.)

So it was that Apollo astronauts were to remain in an all-oxygen, low-pressure environment once in space and throughout the missions.

As the Apollo investigation proceeded, it was inevitable that heads would roll. The most prominent head belonged to Robert Gilruth's second-in-command at the Manned Space Craft Center, Joseph Shea. Shea was Apollo program manager: he'd lived and breathed Apollo; it was his sole focus and passion. But he was fired from his post three months after the fire, in April, and by all accounts he was already disconsolate. It had been his responsibility to oversee and deal with the contractor with all the back-and-forth, changes, and ulcer-inducing complexity that the role guaranteed.

Nearly a year earlier, when the doomed module, CM number twelve had been delivered to NASA during a formal acceptance hand-over ceremony, the three assigned astronauts Grissom, White, and Chafee had had their picture taken with their heads bowed and eyes closed during a prayer of dedication. One of them sent Shea a copy of this photo, inscribed, "It isn't that we don't trust you, Joe, but this time we've decided to go over your head."

For years afterward, Shea kept that photograph prominently placed in his Massachusetts home. By all accounts, he never forgave himself for whatever may have been his responsibility for the fire.

At any rate, and perhaps surprisingly, the giant Saturn V rocket's development—and that of two of the three components that would be vaulted moonward—were not too badly delayed by the fire. Everyone knew that when the newly tweaked Block II Command Module was ready, it would be mated, along with the Service Module and LM, for tests of the whole shebang. Most interim steps involving the smaller Saturn IB would now be hopped over. It was just possible that the necessary tests and procedures could still occur in time to let men walk the moon before the end of 1969.

Of course, everyone was still worried about the Russians.

In 1966 and 1967, there had been increasing signs of Soviet development of a huge new rocket; few doubted that its goal was to place cosmonauts on the moon. Thus a relevant Russian tragedy on April 23, 1967, was no small news for both countries.

It started amid the usual Soviet secrecy, with a launch of the Russian A-2 rocket carrying the new seven-foot-diameter *Soyuz 1* crew module designed to allow three cosmonauts to remain in space for at least ten days. Except, this time there was only a single occupant, the charismatic forty-one-year-old Vladimir Komarov.

It is widely believed that this new Soyuz craft, the first Soviet model that could change its own orbit and perform sophisticated maneuvers in space, was to be the target of a second craft that would be launched to meet it a day later. But that second launch never happened. By all accounts the *Soyuz 1* solar panel deployed in a faulty manner, electrical power was low almost from the start, and *Soyuz 1* suffered serious stabilization and control problems as well. By the fifteenth orbit the situation was critical, and on the seventeenth, Komarov tried to come back.

Unfortunately, he could not orient the craft the right way for a deceleration burn, and had to "go around" yet another orbit to try something else. Western interception of radio communications showed that Komarov spoke to his wife from space, and wept, knowing how badly things were going. He was not really optimistic about a successful outcome.

On the eighteenth orbit, Komarov fired his retrorocket for a ballistic return. Because he apparently lacked the normal stabilization techniques available to fly the gentler aerodynamic return through the atmosphere, his only hope of survival was to spin the craft around its long axis to gyroscopically impart enough stabilization to keep the blunt end with its heat shield aimed downward. But this steeper entry subjected him to twice the usual g-forces during the return, and he had to endure a crushing 10 g's, which frequently produces unconsciousness.

Nobody knows if he was conscious toward the end of this reentry period; what is clear is that when the single main parachute deployed at a height of 20,000 feet, the craft was still spinning, and this spin caused the parachute risers to twist around, squeezing the chute itself into a state of only partial deployment. Like all returning modules the *Soyuz 1* had no other means of reducing speed and slammed into the ground at 350 miles per hour, bursting into flame, and killing Komarov.

Although the Soviets kept this news a secret for much of the next day, they eventually made the announcement, greeted by shock and sadness throughout the country.

American astronauts expressed official and genuine private sorrow as well, and sent words of condolence. Privately, however, U.S. engineers and planners figured that the delay caused by the Apollo fire would now be matched by an equal one in the Soviet Union. The playing field had been leveled once more. (Actually, it wasn't truly level at that point; although they didn't know it yet, in 1967 the Americans were solidly ahead of their competitors.)

As previously mentioned, 1967 was to be the worst in space history in terms of loss of life. No further astronauts or cosmonauts succumbed in space that year, but they certainty did so on the ground. On January 31, just a few days after the Apollo fire, two air force experimenters, William Bartley and Richard Harmon, stepped into an overpressurized pure-oxygen chamber at Brooks Air Force Base to remove biological samples from rabbits that were being subjected to extended periods in the oxygen environment. An electrical short in a piece of equipment caused an immediate raging inferno, and the lives of these trapped men could not be saved.

A series of jet crashes killed additional astronauts on October 5 and November 15, and astronaut Ed Givens died in a car crash on June 6. Also, a high-speed end-over-end tumbling crash of an experimental aircraft on May 10 critically injured test pilot Bruce Peterson, which led to a series of imaginative, revolutionary surgery techniques that inspired the TV show *The Six-Million Dollar Man*. In fact, film footage of his actual crash was shown at the start of each episode.

Although 1967 was the nadir of the Space Race, plans nonetheless went ahead in all aspects of Apollo, especially refinements in the LM's descent and ascent engines, the latter being particularly vital to the lives of the returning astronauts. Then, too, the great Saturn V rocket was finally becoming ready for its critical tests, and was about to replace testing using the much smaller Saturn 1B.

George Mueller, the associate administrator for the Manned Space Flight Program, was perhaps the biggest "brain" behind the souped-up schedule. His plan was to ignore von Braun's plea for incremental testing of the various Saturn V stages using perhaps ten unmanned launches before people were finally allowed to fly. Instead, the testing would be "all-up" meaning everything would be assembled and tested at once, sink or swim, starting at the end of 1967. That could be followed by just one or two more unmanned Saturn V flights, and then astronauts would be packed inside them starting late in 1968, followed by as many as five manned missions in 1969, with the first lunar landing being one of them. At NASA, the Saturn V flights would be designated AS (Apollo Saturn) followed by a "5" and then the mission number: To them, the first Saturn V flight was AS-501. Only the public and press knew it instead as *Apollo 4*. Moreover, merely having *two* redundant and competing number schemes was insufficient: MSF (the Manned Space Flight Center) had their own lettering scheme for the incremental flights that would put people on the moon.

Mission A—this upcoming *Apollo 4* or AS-501 flight—would be the first unmanned flight of the giant rocket and also test the Command Module's heat shield as it returned through the atmosphere at ultrafast speeds.

Mission B was to test the Lunar Module, the LM, which was already starting to look iffy simply because the contractor, Grumman of Long Island, New York, was falling behind schedule. Still, an early version, even if not moon-ready, needed to fly, and a smaller Saturn IB was good enough for this task.

Mission C would be the first manned flight.

Mission D would use both the Command Module (CM or CSM) and Lunar Module (LM) in Earth orbit to see how everything worked together.

Mission E would do the same thing, but now in higher Earth orbit, and check out the higher reentry speeds with people aboard.

Mission F would venture farther, beyond Earth orbit, and maybe even to an orbit around the moon, using all the docking and rendezvous procedures that the actual landing would entail.

Mission G would be the landing on the moon.

Mueller thought that one or more of these might be skipped, but only if everything worked beautifully on each preceding one. But Mission A, or *Apollo 4*, or AS-501 (call it whatever you like) had to work or else everything else would fall far into the future.

This would be the most critical flight of all. Mueller later said, "The whole Apollo program would have gone down the drain if there had been a . . . blow-up on 501. The whole concept of all-up testing was in grave doubt until 501."

Thus it was that on November 9, 1967, von Braun, Mueller, and everyone who had labored for years or decades in the field of rocketry anxiously watched the countdown to this largest-by-far machine, as 4,000 separate parameters were monitored from the control building three miles away. The 700 invited guests and thousands of other well-wishers lined the causeway. The brand-new 3,000-ton machine stood thirty-six stories tall in the rising sun, its million mostly unseen parts now steaming from the ultracold of the cryogenic oxygen and hydrogen fuels.

For the first time, absolutely everything in the countdown was automated during the final 190 seconds. There were simply too many valves to turn on, sequencers to start, motors to begin turning, switches to be thrown, for any human to keep track of them all.

The ignition of the five F-1 engines was dazzling, its light outshining the sun itself. Spectators were so far away, however, that the Saturn was already up past the top of its gantry before the roar began to arrive, its pressure waves pounding everyone's lungs, rattling windows, shaking cars. These were the largest engines in the history of the human race, and their awesome thunderous force made spectators weep with the sheer ultra-subwoofer power of it all. When it was all over, there were no ambiguities. Everything had worked.

Probably the only negative in this flight was one that surprised the engineers: pogo. This is the simple term for a longitudinal oscillation, or expansion and contraction of the rocket while in flight, along its longest dimension. Had anyone been in the Command Module at the top, they would have been terrified at the way the rocket acted like a pogo stick, going faster and slower and faster again in enormous *boings*.

It turned out that the contraction and expansions of the flexible tanks got "in phase" with fuel being pumped to the engines, one exacerbating the other. They were lucky it didn't destroy the craft. Afterwards, engineers worked feverishly and cleverly on the problem, and devised a strategy of slightly reducing engine power just when the rocket was expanding lengthwise, ending the sympathetic resonance. Would it work? The next Saturn V test, another unmanned effort named *Apollo 6*, would tell the tale.

But first there would be a relatively minor (in importance) unmanned flight called *Apollo 5*, the Mission B that did not need to use a Saturn V. It lifted off January 22, using the much smaller two-stage Saturn IB, and was reasonably successful in testing the LM. Then came the second and final scheduled unmanned Saturn V test, AS-502, or *Apollo 6*, on April 4. Without a doubt, this flight was a disappointment.

Things began to go sour very soon after liftoff. Pogo oscillations were so severe that an adapter holding the Command Module broke away a half-minute before the first stage finished firing. The new weight and volume distribution in this flight, set up to mimic actual loads, made the erratic rocket motions worse, not better.

The news only went downhill after this, as the rocket continued upward: The second stage J-2 engines suffered disconcerting failures. Two of them shut down prematurely, with the remaining three unable to even approach the planned high orbit from which the craft would practice Earth entry from the moon. Then, the single J-2 of the Saturn S-IVB third stage refused to turn off and burned for an extra twenty-nine seconds, sending the craft into a wildly elliptical orbit. Worse, when controllers later tried to fire it again, it refused to ignite at all. This was more than a little worrisome. Reignition of the third-stage (S-IVB) J-2 engine was key to TLI, translunar injection, the changing of the astronauts' paths from a simple Earth orbit to a trajectory that would reach the moon. Later, Mueller, exaggerating a little, said, "*Apollo 6* had everything in the world go wrong with it."

What followed next was fascinating. In the normal scheme of things, failures of the magnitude of *Apollo 6* would mean serious troubleshooting, corrections of the problems if they could be identified with certainty, and then another unmanned Saturn V test. In this case, however, the engineers were so sure that they had found the problems and could fix them—serious as they were—that they gave a go-ahead approval for Mission C, the first manned Saturn/Apollo flight.

Really? Put men on the monster rocket on only the third time it was ever flown? With problems that had beset those previous two unmanned test flights? Yes. *Apollo 7* would carry the new Block II CSM that would be hoisted up on an actual Saturn V.

This first true manned shakedown and return to space, using all the key ingredients for the moon flights, would be crewed by the old-man (forty-five-year-old) Mercury and Gemini veteran, perfectionist and Original Seven member, Wally Schirra, supported by newcomers Walter Cunningham and Donn Eisele.

It was one of those events that kept TV sets turned on nonstop around the world. Few would have guessed, however, that it was to get the entire crew fired, and grounded from ever flying again.

◯

# CHAPTER 13

# Insubordination

*The moon like a flower*
*In heaven's high bower,*
*with silent delight*
*Sits and smiles on the night.*
—William Blake, *Songs of Innocence*

I t's a popular theme in sci-fi movies, seen in *I, Robot* and *2001: A Space Odyssey.* The robot or computer is programmed to obey orders, but is also duty-bound to accomplish the mission at all costs. When those two directives conflict, androids get neurotic; smoke comes out their ears and, in fiction at least, they run amok. They're torn: Which command has priority? What should be done?

That was one of the dramatic subplots of the *Apollo 7* spaceflight.

Pilots, particularly test pilots, have strong take-charge dispositions. The "pilot in command" concept endows the person flying any plane with extraordinary privileges. A pilot may ignore or circumvent any Federal Aviation Regulation whenever he or she deems that safety justifies it. A pilot may even perform marriages or arrest an unruly passenger. In the test pilot or astronaut realm, the commander of the ship truly has the last word in ensuring the success and safety of the mission.

Or so it would seem. What about the flight director at Mission Control? Until *Apollo 7*, the tacit sharing of power between ground and space had never come up for dispute. But all that changed when the volatile, no-nonsense Wally Schirra was given command of the first manned Apollo craft after the two-year hiatus following the tragic launchpad fire of mission AS-204.

Schirra had always been a consummate professional, the man you'd want to have in the left seat. But there had been earlier signs during the Mercury years that he just might have a problem with authority—or at least authorities on the ground. During the early sixties, whenever engineers wanted to change something in the spacecraft, Schirra often delivered holier-than-thou cracks such as, "Well, if you'd been *up there* . . . !"

And following the John Glenn flight, Schirra had less-than-flattering things to say about NASA's PR demands; he'd had contempt for the hero-celebrity-lecturer role demanded of returning astronauts, and the role of nonstop diplomat that had been thrust upon Glenn.

But when it came to handling a complex experimental craft, Wally Schirra was good. No, he was better than that: he was almost peerless.

So it was Schirra who was given command of *Apollo 7*, the flight that could put America back in the Space Race. In America's first-ever three-man crew, he would be accompanied by Donn Eisele and Walter Cunningham.

That Schirra was cranky right from the start can be seen in his own appraisal of the launch. Asked to write a short preface to the 2002 book *Apollo* by David Reynolds, Shirra uses the page as a platform to vent complaints thirty-four years after the fact. His feelings about the rocket even being sent up that day:"We weren't supposed to launch if the winds [were high] but they violated that mission rule and launched us anyway.

"When you are given command," he continued testily, "it means you *own* that vehicle. . . . They usurped my command."

He finishes up the diatribe with, "I've never heard of a flight controller risking his life by falling out of a chair."

Of course, on October 11, 1968, the world was unaware of those dour feelings. The two-stage Saturn IB launch—the first ever with a manned crew—went perfectly and the orbit was right on target, its apogee only two miles short of the intended point. After the far smaller Atlas that hoisted Mercury missions aloft, and the Titans used for Gemini, this new rocket was a colossus even if it wasn't yet the true behemoth, the Saturn V that would hurl people skyward soon enough.

The crew settled down to what would be eleven days of tasks that had been meticulously scheduled during arduous months of flight planning. Refirable engines were tested, equipment checked out, orbits changed, all done by Schirra and company to perfection. This Command Module was like a luxury suite compared with the claustrophobia-inducing Mercury and Gemini capsules. With Apollo, especially with the center seat stowed, there was actually a little room to move around. If one stood in the lower equipment bay, one's head wouldn't even touch the ceiling.

The good feelings were not to last. The first day in orbit had not even ended when first Schirra, and then the others, came down with serious head colds. Sinuses were clogged, and in a weightless environment this meant that the nose wouldn't run because there was no "down" to run to. Instead the cavities would remain stuffed. A relatively new medicine, Actifed, was fortunately in the medical kit, and the crew gobbled these for the entire flight. No wonder Schirra later enthusiastically accepted an offer to endorse the product in commercials.

Trouble also arose when NASA's PR department sent up requests for the astronauts to perform in front of their new miniature onboard cameras. This was the first time that a large cabin, mobile crew, and improved cameras would allow the world to follow along in the action, and Mission Control asked the crew to say a few things to viewers below, a kind of live TV show.

Schirra would have none of it. "The show is off!" he snapped. "The television is delayed without further discussion. We've not eaten. I have a cold. And I refuse to foul up my time."

Lead Flight Director Gene Krantz could scarcely believe his ears. Until now, no spaceflight had ever dealt with such insubordination. Schirra did finally schedule a TV show a day later, and a good one too. "Hello from the lovely Apollo room high above everything!" he chirped; and the crew did several more during the eleven-day mission. As for the insubordination, however, it was just the beginning.

The main gripe, as the crew saw it, was that they had a busy and intricate work schedule whose purpose was nothing less than checking out a host of new systems and procedures that they hoped would enable a moon flight the very next year. When sudden, new (and to their minds, unnecessary) experiments were radioed up to them, they were astonished. Of course, engineers on the ground simply wanted to demonstrate that the spacecraft could perform in every possible way should the need arise. But Schirra would sometimes refuse:

"We have a feeling," he radioed down, "that some of those experimenters are holier than God down there. Well, we are a heck of a lot closer to him right now."

Cunningham echoed his commander's sentiments: "People who dream up procedures like this after liftoff have somehow or other been dropping the ball the last three years!"

Tensions only got worse, and Schirra finally barked, "I have had it up to here today, and from now on *I'm* going to be an onboard flight director for these revised schedules. We're not going to accept any new games—or do some crazy tests we never heard of before!"

Every test pilot who ever had to buck a bureaucracy probably quietly cheered Schirra's stick-it-to-them tenacity. But the flight director was amazed. Later, Gene Krantz said, "He was sure ornery. I don't think that anyone has ever figured out what it was that was really the burr under his saddle." And Mission Control's all-powerful head, Chris Kraft, assured colleagues around him that neither Schirra nor any of the other crew members would ever fly again.

With Schirra it hardly mattered since he'd already announced that he was quitting the astronaut corps after this flight anyway. But for Cunningham and Eisele, this shakedown flight effectively canceled their tickets to the moon. Other astronauts were rotated into actual landing missions, but not them. Kraft kept his word. They never flew on Apollo again.

For the moment, of course, they had a spacecraft to handle, and this they did flawlessly. And, in Schirra's defense, even Krantz later admitted that, "We kept piling a lot of stuff on him, because with only one test to get the job done, every time we saw an opportunity we'd go for it. We'd make engine burns longer than he'd seen before in the pre-mission planning."

You can't blame them. Engineers and planners knew that many problems would only show up in space. Surprises would be revealed. New electrical generators that worked one way on Earth would work differently in the vacuum of space.

Also, the world didn't know it yet because it was being kept secret, but the decision had already been made that if *Apollo 7* was completely successful, *Apollo 8*, a mere two months later, would not merely be another orbital system checkout—it would actually go to the moon. Schirra's critical, golden, and ultimately successful mission ensured that the United States would indeed orbit astronauts around the moon ahead of the Russians.

Now it was time to return, and there was one last confrontation to be played out with ground controllers. Schirra and his men still had stuffed heads, and he wanted to ride back to Earth with the helmets off, to help clear themselves when the pressure changed. Deke Slayton on the ground, trusted by all the astronauts because, after all, he was one of them, urged Schirra to follow procedures and to please wear the helmets. The logic was impeccable: No manned command modules had ever returned from space.

What if a leak suddenly developed during the fiery stress-filled, high-speed return? Only the pressurized suits would save them.

Schirra would have none of it. He was commander, and he announced that he was making the final decision to disobey orders one last time. The helmets remained off.

Three years later, Schirra's judgment was proved wrong. In June 1971, the *Soyuz* spacecraft was performing reentry just like *Apollo 7*, when the separation charges fired together instead of in sequence, and the additional shock popped open a vent. Despite frantic efforts to manually close the valve, the *Soyuz* interior was a vacuum in less than a minute. There simply was not time enough for the Russian crew to don their suits, and the three asphyxiated. Of course, that doesn't imply that the same thing would have necessarily befallen their American colleagues, but, still. . . .

Thirty years later, Schirra thought back to his astronaut days, and had no regrets about not flying again: "I went up for a long time and I found it extremely boring just orbiting Earth for eleven days."

But it was anything but boring for NASA: in fall 1968, Schirra's first manned Apollo had been "105 percent successful." Three astronauts, in beards, stepped aboard the aircraft carrier USS *Essex* in the rain. And just like that, the moon no longer seemed so far away.

○

# CHAPTER 14

# First to the Moon

*When the moon hits your eye*
*like a big pizza pie—*
*that's amore . . .*

Harry Warren and Jack Brooks, "That's Amore"

History is rarely kind to those who accomplish great things but then find their feats surpassed. We remember the men who first climbed Mount Everest, but not those who conquered Pike's Peak or Mont Blanc. And so it was that in 1968, three astronauts got to be the first life forms to break free of Earth's gravity, first to ride the mighty Saturn V rocket, first to go to the moon—and yet are remembered today by virtually no one.

Does the name Bill Anders ring any bells? Probably not. But Anders and Jim Lovell and Frank Borman looked back on the whole Earth and became the first people ever to see entire continents, the first to see our planet as a sphere floating in the void.

## How Does Earth *Really* Look?

Astronauts on the Shuttle, or fixing Hubble, or aboard the International Space Station, are commonly thought to have "escaped Earth's gravity" and to be "in outer space." Using TV and hand-held cameras with ultra-wide-angle lenses, they transmit images of Earth that shows a wildly curved horizon.

It's all a form of fakery.

Astronauts orbit less than 250 miles up. This is just one-thirtieth the diameter of Earth. They are barely above the atmosphere. The horizon does show a curve, but it's a slight curve; it's almost flat. The strongly curved televised limb is an artifact of the special wide-angle or fisheye lenses. Shapes of land seen below them in such telecasts are merely islands or peninsulas; they are never entire continents, which is what one would view in a true, undistorted image of a complete round Earth.

By contrast, the moonbound astronauts aboard *Apollo 8* and then *Apollo 10* through *17* went 300 times farther than anyone else before them. They alone saw the Earth as a complete sphere, pole to pole. They alone could apprehend entire continents. To them alone did our planet appear as a two-degree ball dwarfed by the surrounding space. They alone beheld Earth in its truest perspective—and were awed by the experience.

The mission itself came about secretly, as a surprise. In the summer of 1968, before anyone had flown a single manned Apollo test, several key NASA planners discussed the idea of jumping ahead to send the second manned mission all the way around the moon instead of merely testing systems in Earth orbit, as the first manned mission, *Apollo 7*, would do.

Their reasoning was born of paranoia. The Russians were known to be rapidly developing the capability of sending people around the moon. Any time now, the Soviets could stuff a couple of men inside a crude Soyuz and then rightly claim to be the first country to have gone to the moon. No matter that they didn't land—it would take some of the wind out of America's sails.

The United States wanted to do it first, and could, with a quick game of cosmic three-card monte. The official launch schedule called for a series of Earth-orbital tests. This made sense because the LM, the Lunar Module, wouldn't even be ready to fly until early 1969. But if the first shakedown flight was successful in October, why not send the next crew around the moon *without* the lander?

In August, this proposal started circulating among NASA's heavy hitters, and everyone loved it: von Braun was on board and so was Gilruth, Mueller, Paine, and Petrone. Now they only had to sell it to NASA director Jim Webb, who was in Vienna attending an international space conference.

It went over like a lead balloon. Webb didn't like the idea at all. In fact he was furious when he realized that everyone had been hatching this plot in his absence. He was particularly uncomfortable with the idea of twisting the carefully sequenced series of Apollo tests. Nonetheless, he ultimately came around, but insisted on one proviso: the true destination of *Apollo 8* would be kept from the press until after *Apollo 7* returned. If *7* was a total success, then he'd okay *8* for the moon.

And this is how NASA grabbed a page of one-upmanship from Russia's Cyrillic alphabet. Their team's agenda: Let's beat the Russkies at their own game.

Apollo 7 was a success, and NASA immediately released the news of the dramatic upcoming change, to the astonishment of all. *Apollo 8* would be heading for the moon.

The wisdom of the decision was quickly apparent. On September 12, the Russians launched their unmanned *Zond 5* craft, and three days later it looped around the backside of the moon and returned to Earth. It took the easy ballistic route into Earth's atmosphere and parachuted some animals gently down after subjecting them to a crushing 15 g's of deceleration. Then, a month after *Apollo 7* returned, on November 11, Russia sent *Zond 6* around the moon, and this time performed a sophisticated aerodynamic (lifting) entry that gave it a much gentler atmospheric penetration. ("Zond" was a meaningless code name Russia often used for test launches, when they didn't want anyone to know what they were up to. When earlier Zonds failed, they didn't have to answer to the world because they'd never announced what the flight was supposed to do in the first place.)

At any rate, it was clear that Russia could now shoot people *around* the moon at any time. The December date for *Apollo 8* might happen barely in time . . . and maybe not.

Anders, Lovell, and Borman stepped into their 214-cubic-foot module at dawn on the equinox, December 21, 1968. Crowds lined the causeway at Cape Kennedy to see this new amazing rocket, and they were not disappointed. At 7:51 a.m., the five giant F-1 engines—first conceived in 1953 and finally built only in the mid-1960s—burst into its deafening roar, the loudest sound ever created by humans, outside of an H-bomb blast. Three thousand pounds of kerosene sprayed together with liquid oxygen each and every second, and the six-million-pound rocket rose slowly upward.

After two minutes and forty-one seconds, the awesome first stage shut down, with the Saturn forty-two miles high. The five J-2 engines of the second stage then ignited with their distinct, odd, clear bluish flame, while spectators gasped and cheered. The crowd could also see the now-unneeded escape tower fall away, along with the cylindrical staging ring that had held the first two stages together.

The astronauts felt these new second-stage high-tech engines accelerate the craft through 17,000 miles per hour in a jerky six-minute ride that was unpleasant and disquieting. Previous Saturn flights had measured this uneven effect, a pronounced faster-slower-faster acceleration as if the engines were sputtering. The astronauts were glad when this pogo stage had finished its job and the single third-stage (S-IVB) J-2 engine took over. Interestingly, this would fire for the same two minutes, forty seconds as the first stage, and then they were in orbit.

The crew would spend nearly three orbits, or two-and-a-half hours, checking all systems, before getting the okay for the TLI, the trans-lunar injection. This was the burn that would add 6,600 miles per hour to push them free of Earth and start them on their three-day glide to the moon.

But first this single J-2 engine had to reignite. Everyone remembered that on the most recent Saturn V test only a few months earlier, this engine had refused to light. So it was with no small sigh of relief when, now, it did. The crew was now off, and humans were leaving Earth behind for the first time in history.

After engine shutdown the astronauts separated their command module from the spent third stage, and began their three-day coast. For the next fifty-three hours of the trip, Earth would be tugging at them, trying to bring them back, slowing their craft from their present speed of 24,000 miles per hour down to an anemic soap-box-derby 2,500 miles per hour. At that unmarked point, empty of signposts, they would become the first humans to reach the place where Earth and moon gravities meet each other. After this it was literally all downhill: now they'd start accelerating again.

This was a major safety feature of the entire plan. If their Service Module (SPS) engine failed to fire, their natural free-falling trajectory would loop them around the moon and throw them automatically back to Earth, a zero-energy ride whose small tweaks could be accomplished with just the CM's little thrusters.

Like passengers in a train's backward-facing seats, the crew could not see where they were going. Oriented tail first so that their SPS engine could apply braking at the right moment, Borman, Lovell, and Anders had no glimpses of the moon's slowly increasing size as the hours and days went by. It was only after their engine fired and they actually entered orbit that the breathtaking panorama suddenly unfolded outside their windows.

Now came an important decision. After carefully checking their trajectory and all systems, Mission Control could give them the go-ahead for the SPS engine firing that would place them into lunar orbit. Once that had been done—and they had to fire the engine when the ship was on the far side of the moon, out of radio contact—the decision would be irrevocable. Then they simply could not get home unless this same SPS engine later fired once more, to add the same 2,400 miles per hour they would now subtract. Firing it once meant that it must ignite twice.

They got the go-ahead barely minutes before they vanished behind the moon. Now everyone back in Houston could only wait. If the engine failed to fire, they'd emerge from the moon's opposite limb twenty-three minutes later. If all went well, and their four-minute, seven-second burn had occurred as planned, their reappearance would be delayed and they wouldn't appear for thirty-three minutes. It was a rare case when being late was a good thing, when no news would be good news.

For the crew, this time spent utterly alone would later be described by various later missions as particularly attractive. Here, for the only time, no one back at Houston could bother them. Their curtains were closed, they had rare privacy, they were men in the middle of a vast sea out of sight of land. On this particular flight, they became the first humans in the history of the solar system who could see no trace of planet Earth, no matter where they looked. How that must have felt is something none of the rest of us can ever know.

They gained another little-appreciated merit badge as well. At this point in the mission, they gained the distinction—one curiously never mentioned in later press accounts—to be the first human beings ever to gaze upon the moon's far side.

This should not be confused with the "night side" or "dark side," both of which are always temporary. The full moon always displays the same features to us, with just a slight wobble called libration that makes dark blotches near the edges shift position a bit. Even the most ancient cultures noticed this strange and unique lunar property, that the same face is forever aimed in our direction—and it remains a singular quality even in our modern times of powerful telescopic exploration. There simply is no other body in the universe that forever shows us just one of its sides. (Mercury comes closest, by exposing the same hemisphere to us on every alternate revolution around the sun.)

The flip side of this curious coin is that there exists a lunar hemisphere forever unseen from Earth. Until October 4, 1959, when the Russian spacecraft *Luna 3* flew past the back side and transmitted its initial low-definition images, all we could do was guess how it would appear. Virtually everyone assumed that the far hemisphere would look a lot like the one we see.

Wrong. Turns out, the distant hemisphere appears entirely different, with far fewer extensive dark areas. On the near side, many cold, frozen, ancient flows of lava have drowned out myriad even-older craters, and created enormous, dark, smooth zones that are the most conspicuous features visible to the naked eye. The far side almost completely lacks these maria, or seas. Instead it displays all the many cratered scars it received during its first few hundred million years of life; the back side is much more rugged than the side we see. The two faces look like different worlds. (It might be argued that Earth, too, has two very different hemispheres, since the Pacific basin with its vast oceanic panorama scarcely resembles the opposite, continent-filled face).

Because radio communication with Earth is easily blocked by the moon itself, all of the Apollo missions performed the actual landings on the near side, and we still have no hands-on samples from the distant hemisphere. Again it must be stressed that both lunar hemispheres receive periods of day and night. Neither is permanently dark. (Reality is unlikely to deter future writers, who apparently find attractive the notion of a world with one side forever in shadow. Pink Floyd's famous album *Dark Side of the Moon* will surely not be the last we'll hear of that mythical, Oz-like—but nonexistent—realm.)

Now, however, the Apollo crew gazed in radio silence at this most truly alien landscape that had only been photographed by robotic orbiting probes, and never seen directly.

Thirty-three minutes later, as they passed over the unmarked boundary to the familiar lunar near side, they suddenly found themselves back in radio contact, in a free-falling elliptical path around the moon. After two full lunar circuits the SPS engine was fired again briefly, for just nine seconds, and this

Herman Oberth (1894–1989), one of the three widely recognized founders of rocketry, was the only one alive to attend the launch of *Apollo 11*.

Photo courtesy of NASA

Mission control in 1926. Robert Goddard launches a rocket from the safety of his control building. Sandbags have been added to fortify the roof.

Photo courtesy of NASA

Same day, different angle, as Goddard holds the crude controls. He could launch or abort—it was that simple.

Photo courtesy of NASA

Goddard's most famous photo, his best rocket at the time, which used liquid oxygen and gasoline propellant. Taken before launch on March 16, 1926 at a relative's farm in Auburn, Massachusetts.

Photo courtesy of NASA

Goddard's double-acting engine, with each pump operating independently, was one of his technological breakthroughs. Ten years later his more advanced designs would ascend 8,000 feet.

Photo courtesy of NASA.

Hermann Oberth (forefront) with officials of the Army Ballistic Missile Agency at Huntsville, Alabama, in 1956. Left to right: Dr. Ernst Stuhlinger (seated); Major General H. N. Toftoy, Commanding Officer and person responsible for "Project Paperclip," which took scientists and engineers out of Germany after World War II to design rockets for American military use. Many of the scientists later helped to design the Saturn V rocket that took the Apollo 11 astronauts to the Moon. Dr. Eberhard Rees, Deputy Director, Development Operations Division Wernher von Braun, Director, Development Operations Division.

Photo courtesy of NASA

Mercury technicians prepare
a capsule to send a single man
into orbit.
Photo courtesy of NASA

Not much room inside,
but plenty of glory for
the brave members of
the "Mercury Seven"
astronauts, who crammed
into these tiny capsules
for their brief trip above
Earth's atmosphere.
Photo courtesy of NASA

Mercury astronauts rode atop dangerous, anemic, alcohol-and-liquid-oxygen fueled Redstone rockets.

Photo courtesy of NASA

*Gemini 3* is mated to its *Titan II* launch vehicle—a modified ICBM. Two men spent up to 14 days crammed into its claustrophobic space barely larger than a phone booth. They were unable to get up or even extend their legs, and they had to remain in their original clothing the whole time.

Photo courtesy of NASA

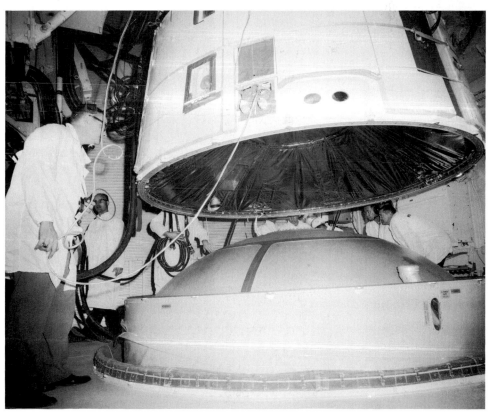

No astronaut had ever "nailed" a spacewalk until 1966, when Buzz Aldrin showed the world how it should be done. His amazing performance on *Gemini 12* made him deserve his place on the first team to touch the moon.

Photo courtesy of NASA

The world's largest rocket rolls slowly away from the world's largest building, the VAB, where it was assembled. The same scene had first unfolded in a fictional silent movie nearly a half century earlier.

Photo courtesy of NASA

Two seconds after ignition, the massive F-1 engines are already consuming 3,000 pounds of fuel per second, as a *Saturn V* heads upward toward the moon.

Photo courtesy of NASA

*Apollo 1's* interior. The three astronauts didn't have a chance: A fire was all but inevitable, and the hatch simply couldn't be opened from the inside.

Photo courtesy of NASA

*Apollo 10* astronauts came within 10 miles of the surface but didn't land. As they circled the moon, they took amazing photos of its gray airless surface.
Photo courtesy of NASA

Neil Armstrong, Michael Collins, and Buzz Aldrin in a rare, informal, truly relaxed photo— before their names become known around the world.
Photo courtesy of NASA

The most famous footprint in human history, created at 10:56 P.M., Eastern Daylight Time, on July 20, 1969. The surface was covered with thick fine power, but just a few inches down it became so tightly packed, digging was nearly impossible.

Photo courtesy of NASA

Earth, photographed by the first people to touch the surface of another world.

Photo courtesy of NASA

A captive audience. President Nixon used every opportunity to appear in Apollo spotlight. Here he meets (L to R) Armstrong, Collins, and Aldrin in their temporary quarantine chamber aboard the USS *Hornet*, after their return. He'd had the Presidential Seal attached to the door, in sight of the TV cameras but invisible to the astronauts.

Photo courtesy of NASA

Sunrise at the Sea of Tranquility, which is the smooth dark area on the terminator (shadow line) at the top. *Apollo 11* was three days into the mission at this point, and it landed a day later.

Photo courtesy of the author

The near side, where all the Apollos landed, is much smoother. This photo shows the moon's phase, and lighting conditions, as *Apollo 11* left the moon from its landing site at the Sea of Tranquility, marked with a black dot

Photo courtesy of the author

*Apollo 12's* remarkably precise landing let the astronauts walk just 600 feet to the unmanned *Surveyor III*, which had landed 2 ½ years earlier. No one yet suspected that it might harbor an astonishing surprise.

Photo courtesy of NASA

James Irwin readies the Lunar Roving Vehicle for its very first use, during Apollo 15. In the background is 14,765-foot-tall Mount Hadley, whose dimensions and distance are impossible to judge in the moon's stark, strange, hazeless lighting.

Photo courtesy of NASA

Harrison Schmitt, the only moon-walking scientist, spots some odd orange soil during Apollo 17, the final mission to the moon. Apollos 18–20 had already been cancelled.

Photo courtesy of NASA

HERE MAN COMPLETED HIS FIRST
EXPLORATIONS OF THE MOON
DECEMBER 1972, A.D.
MAY THE SPIRIT OF PEACE IN WHICH WE CAME
BE REFLECTED IN THE LIVES OF ALL MANKIND

EUGENE A. CERNAN
ASTRONAUT

RONALD E. EVANS
ASTRONAUT

HARRISON H. SCHMITT
ASTRONAUT

RICHARD NIXON
PRESIDENT, UNITED STATES OF AMERICA

Final plaque left on the moon. Like the others, it was attached to one of the legs of the LM. Only the name Richard Nixon repeats at all six landing sites. The words should remain legible for at least a million years.

Photo courtesy of NASA

The most famous photo of Earth ever taken, this image was recorded during the final mission to the moon, Apollo 17. This was the only time that any Apollo astronauts saw a "full Earth."

Photo courtesy of NASA

Where will rocketry go next? In this image, all eight planets, plus the moon, have been photographed not through earthly telescopes but from visiting spacecraft that had arrived at each in turn. The rocket's next major new role will probably be a mission to transport people to planet number four, whose true color is not red, but tan.

Photo courtesy of NASA

Total solar eclipse. The moon and sun are the only round objects visible in the sky to the naked eye. Here one fits perfectly over the other, adding to the moon's historical mystique.

Photo courtesy of the author

The far side is much more heavily cratered than the hemisphere visible from Earth. To this day it has only been seen by 27 people, the crews of Apollo 8, and 10–17.

Photo courtesy of NASA

final tweak circularized their orbit to an unvarying sixty-nine-mile height above the surface.

At 3,640 miles an hour, they were orbiting five times slower and they were three times closer to the ground than a comparable Earth orbit, and they drank in the amazing views of the gray surface. (This was closer to the ground than astronauts can ever orbit around Earth because a sixty-nine-mile-high path would be quickly degraded by the drag from the thin air of the upper atmosphere.) They were astonishingly, breathtakingly close to the moon. It was Christmas Eve, and people around the world listened to radios to hear the first live accounts of the moon's appearance.

As with all future Apollo missions, each two-hour orbit around the moon had several distinct visual components. Part of the time the crew flew over sections of the moon that were visible from Earth, and that therefore contained features known by every serious amateur astronomer. During the other half of each orbit they glided in radio silence over the rugged far side.

These hemispheric halves were further subdivided into three distinct visual experiences based on the lighting below them. One, of course, was simple daylight, the glaring white sun shining with full intensity on the lunar terrain. Another extended period found the moon unlit by the sun but nonetheless still receiving earthlight, the bright reflected radiance from our world. When earthlight was present the moon's dark side still had a bright glow and features were clearly seen. For the astronauts, the Earth itself was brilliant in the sky, some fifty times brighter than the full moon appears in Earth's sky.

But when the men were over the slice of moon where both sun and earthlight were absent—this always happened of course during a period when they were on the far side—then for the only time in their mission they knew true and utter blackness. The stars would suddenly burst alive in the sky in all their splendor, as if someone had flicked on a switch.

It was Christmas Eve, and Commander Borman asked to read a prayer that he had intended to deliver to the congregation of his Episcopalian church, but hadn't because, "I couldn't quite make it." Permission was granted, and those listening heard him intone: "Give us, O God, the vision that can see thy love in the world in spite of human failure. Give us the faith to trust thy goodness in spite of our ignorance and weakness. Give us the knowledge that we may continue to pray with understanding hearts. And show us what each one of us can do to set forward the coming of the day of universal peace. Amen."

*Apollo 8*'s success was total and sweeping. The SPS engine fired successfully, they glided home for three days, and hit their reentry point to perfection. Because of the 25,000-mile reentry speed, it was a more dangerous and taxing return to Earth than any that had ever been experienced before. In order to minimize the g-forces, a clever mathematical deceleration had been planned, that would be followed by all future Apollo returns as well.

First, they had to strike Earth's atmosphere almost horizontally, at the very slight downward angle of just six degrees. This was critical: a mere two degrees lower and they would burn up like a meteor; just two degrees higher and they would bounce off the air like a flat stone skipping along a lake surface. They would skip completely off Earth's atmosphere back into space and never be able to return.

This was called a "lifting entry" because the trajectory used the flat bottom of the CM, plus the shallow angle, to skim gently into the atmosphere down to an altitude of thirty-four miles, where their momentum in the thicker air would actually "bounce" them back upward to forty miles, before falling once again. It would keep the forces within the module to within 7 g's, which for a test pilot is almost a child's carnival ride. From first atmospheric penetration until final splash into the sea, the craft would travel 1,500 miles around the world.

It all went perfectly, which was particularly nice since this was the first-ever night landing, with its inherent limitations of vision for the rescue helicopters. The astronauts walked wobbly-legged aboard the aircraft carrier USS *Yorktown* just ninety minutes later.

As for the Russians, along with their congratulations they quickly issued a statement saying that they never intended to send people toward the moon in the first place. They said they were concentrating their space efforts on such implicitly superior goals as exploring other worlds with automated probes. Translation: You beat us, and we're not even going to go to the moon now, rather than be second.

And just like that, on December 27, 1968, the United States had won the Space Race.

On January 10, 1969, New York toasted the returning heroes with its wild parade and blizzard of confetti, and it was Chicago's turn the next day. Two weeks later, Europe had a chance to cheer the astronauts, during their nine-country tour. The Russians had been beaten. It was all over.

○

# CHAPTER 15

# Voyage to Oblivion

*And there's the moon.*
*There is something faithful and mad.*

—e. e. cummings, "Is Five"

The new year of 1969 brought optimism to NASA and the American public alike. But it was tempered with a nervous caution. Many suspected that recent Soviet announcements of disinterest in sending men to the moon were designed to throw the world off guard, that they'd still manage to land before the summer, that they'd stay true to previous form and somehow beat the Americans like the Road Runner outsmarting Wile E. Coyote.

But those in the know thought this unlikely now. The United States had just two more preparatory flights and then the real thing would come in mid- to late July when the moon was correctly positioned.

The first of 1969's two preparatory flights was *Apollo 9*, and if ever there was a bad Apollo assignment, this was it. NASA knew that this first-ever shakedown flight of the LM in Earth orbit was critical to the moon landing. Moreover, the docking and undocking procedures, and the practice spacewalk in case something went wrong during an actual moonflight, would make this *more* demanding than an actual trip to the moon. No matter. It all was to be done in Earth orbit, and to the press and public it seemed like a giant anticlimactic step backward. We'd already gone around the moon. Now the next Apollo would only be orbiting Earth? Ho-hum, and wake us up when you decide to do something exciting.

The unlucky astronauts for this yawner of a mission were Rusty Schweickart, Jim McDivitt, and Dave Scott. They had been rehearsing for this flight since 1966, and in a way, flying the LM for the very first time should theoretically have been every astronaut's dream. Of course, in the original planning this mission wasn't supposed to be flown *after* another team had already gone to the moon and been celebrated in ticker-tape parades.

The change of sequence was not just a public relations bust, it had another undesirable effect on the crew: with crew rotation schedules, they now wouldn't likely be flying again until *Apollo 18* or *19* or *20*. Unbeknownst to everyone at this point, those final three flights to the moon would be canceled. Result: Of the capable *Apollo 9* crew, only Dave Scott ever made it to the moon (on *Apollo 15*, two years later). The other men were not destined to ever go anywhere near it.

## What's That Thing Called?

The spidery-looking Lunar Module was the centerpiece of the Apollo missions: it was the *only* machine that would actually make contact with an extraterrestrial body. What was it called? At first NASA designated it the Lunar Excursion Module, or LEM (pronounced "lem") for short. Then NASA's PR people decided that the word "excursion" sounded too picniclike and casual for such a dangerous and historic enterprise, so the word was dropped and the complex flying machine officially and permanently became the LM. Nobody, however, is able to pronounce "LM" monosyllabically so everyone in NASA and in the press kept calling it the LEM anyway.

While we're at it, don't forget that the Command Module, or CM, was often designated CSM because it spent virtually all of its time in space connected to its Service Module, which supplied rocket power and a dozen other necessities. CM, CSM, LM, LEM—this was only the iceberg's tip when it came to NASA's endless abbreviations and acronyms.

The press and public duly ignored *Apollo 9*. After it landed, there were to be no screaming crowds, no trips through Europe. Schweickart, Young, and McDivitt never got their fifteen minutes of fame. But they put in far more than their fifteen minutes of hard work. Their job was to make certain that the LM was spaceworthy, and to determine if docking and undocking could be accomplished. It was a far more demanding assignment than *Apollo 8*'s romp around the moon two months earlier.

The ride began on March 3, 1969, with the ear-splitting low thunder of the giant Saturn V, standing twice as high as Niagara Falls. After a typical smooth ride on Saturn V's behemoth first stage, the crew again endured a jerky pogo experience on the S-II second stage. As before, the third stage ended the anxiety and was a pleasure, and the men were in orbit.

The crew's most important initial job was to try out the docking probe for the first time; this was its inaugural flight on the CSM. This all-important mechanism was designed to firmly connect the LM and CSM. It employed an extending device that would push into a circular hole in the center of a cone on the LM. The navy men in particular were very familiar with this basic concept.

Once mated momentarily, three latches on the docking ring would grab the inside of the circular flange. These initial latches would provide the "soft dock"; then nitrogen gas would fire twelve more latches to hold things rock steady—the final "hard dock." In practice, the two ships could be firmly mated even if just three of the twelve were holding. Seals made of silicon would compress to provide an airtight fit when the ships were connected. The seal would be so firm that men could transfer from one craft to the other wearing nothing but T-shirts.

It was Scott who controlled the joystick to ease the CM into a 180-degree flip-over turn. The Lunar Module had been stored in its cylindrical "garage" atop the Saturn third stage, and now, on command, this housing fell apart into four petals, revealing the LM. Scott carefully pulled closer to link up with it. When contact was made, and the twelve latches explosively fired on cue, Scott gleefully announced, "We have hard dock."

On hitting a control that remotely communicated with the third stage, powerful springs pushed the two linked spacecraft away. The crew now gingerly fired small thrusters on the third stage to move it out of their way, and when it was a half mile in the distance, they remotely fired its main engine, sending it safely into a permanent orbit around the sun. The discarded third stages of all the Apollo missions would remain miniature planets orbiting the sun forever, except for those deliberately aimed for a crash into the moon to measure subground vibrations using the seismometers left on the surface by the various landing teams.

## The LM

Standing as tall as a two-story house and weighing 8,500 pounds, the Lunar Module took Chief Engineer Tom Kelly's 3,000-person team six years to create and perfect. Because it would be used exclusively in a vacuum, it had no need to be streamlined. But saving weight was critical; so important, in fact, that everything was utterly minimized. The skin of the craft was only as thick as three sheets of aluminum foil in places. The legs used honeycombed compressible aluminum instead of normal shock absorbers for its single experience of slamming onto the lunar surface. Space restrictions made its crew compartment scarcely larger than a closet. To save weight, seats were abandoned; the astronauts would merely stand at their control posts. Astronauts would have to spend up to three days in this cramped space, where there was only marginally enough room to get in and out of their space suits. Yet the LM could maintain a comfortable seventy-five-degree temperature despite being on the broiling lunar surface. It had its own radar, and an ability to communicate with Earth as well as with the CSM. And, of course, its ascent engine was built with simplicity and reliability in mind. It had to fire on each mission. And each time, it did.

Among the accomplishments of *Apollo 9* was a full in-space tryout of the lunar space suit. It had a four-hour oxygen supply and another thirty-minute backup oxygen unit in case the primary failed. Another successful task was putting the LM through an endless series of motions, turns, dockings, and, ultimately, a long and distant separation from the CM with McDivitt and Schweickart inside, piloting the LM for more than six hours.

When they returned after ten days, George Mueller described *Apollo 9* as one of the most successful missions ever flown. He was right. Of *Apollo 8, 9*, and *10*, there is little doubt that the goal accomplishment of *9* was the most critical for the moon landing. Yet this was the mission that history would most ignore.

Only a single additional test was needed before an actual moon landing could be attempted. This was *Apollo 10*, which would be a teaser. The plan called for a full landing mission in every way, including the release of the manned LM for a descent toward the surface. *But*: The LM would stop short of the lunar terrain by a mere 45,000 feet, about the height that a jet airliner flies. Then it would discard its descent stage, fire its ascent stage, and rendezvous and dock with the mother ship, the Command Module. They would then fire their single engine and rocket out of moon orbit and head back home.

When we heard this plan back in 1969, many of us slapped our heads. Going 99.99 percent of the way and *not* landing? Why not? George Mueller himself argued in vain that venturing so close with *Apollo 10*—Mission F—was an unnecessary test that could be omitted, and that *10* should go all the way and land. Many held out that hope that NASA was withholding a secret surprise, something to fool the Russians, and that *Apollo 10* would decide to land after all, in the last minute. Or maybe, some whispered, the crew would take it unto themselves to keep going. Little did the public know that NASA planners had already made this impossible. The LM's ascent stage was fueled to just 63-percent capacity. Any mischievous temptation had been nipped in the bud; they *couldn't* make it all the way back up from the surface.

The flight blasted off on May 18, carrying astronauts Tom Stafford, John Young, and Gene Cernan. The story was the same as before. Although NASA was trying what they could to smooth out the bumpy second stage with a routine that had the middle J-2 engine shut early (the outer four engines would continue firing for an extra minute to compensate), it didn't help at all. In fact, this trip into space was the most turbulent of all.

After the smooth two minutes, forty seconds of the ultrapowerful first stage, whose job was to throw the six-million-pound, thirty-story-tall machine to a height of forty miles, the S-II second stage began its customary rough ride. The pogo effect was accompanied by such bone-rattling shaking that the crew was concerned that something was wrong. When the S-IVB third stage fired, it didn't produce its usual relief. Its single J-2 engine instead bounced and vibrated so much during its own two-and-a-half-minute firing that John Young jocularly asked Mission Control if the LM hadn't fallen off during the flight.

The brief checkout during Earth orbit, the extraction of the LM from its garage, and the SPS firing for the moon trajectory all went flawlessly, and the crew decided they had plenty of time to break out the TV cameras. These were the most ham-oriented astronauts to date, and they delighted viewers around the world with numerous shows from the space near Earth, and then from the moon's vicinity.

NASA meanwhile had decided to once again allow semiofficial names for their spacecraft. It had been a practice during Mercury (*Freedom 7*, for example), then was disallowed after Grissom's early Gemini flight (*Molly Brown*), when modules were saddled with bureaucrat-pleasing letter-and-number designations. Now, with Apollo, NASA decided to loosen up. *Apollo 10's* Command Module was *Charlie Brown*, while the LM was *Snoopy*. No doubt echoing the public mood, media announcers and reporters loved it.

This mission was being flown under the identical lighting conditions that would be found in July, and one of the objectives was to go down for a close look and see if the primary landing site at the Sea of Tranquillity was really as smooth and boulder-free as it seemed from all of the previous photographic reconnaissance. They'd also be checking out alternative landing places, all in preparation for the real thing.

When, during their second day out, the crew was told that their trajectory was so perfect that no more engine burns were required, they could just coast to the moon, Young radioed back, "Looks like it'll be cheaper to keep going than turn back, huh?"

On the fourth day, the crew fired their big SPS engine for fully five minutes, fifty seconds in order to lose 1,800 miles per hour and place them in orbit around the moon. With the LM, they were fifteen tons heavier than *Apollo 8's* visit to the moon, and it required an extra minute of burn to slow down this greater mass.

A further burn circularized the orbit a mere sixty-nine miles above the surface, and now they were "home" so to speak, and shared the amazing view with a half-hour TV broadcast out the window.

Gene Cernan and Tom Stafford crawled through the tunnel, spent hours meticulously checking out *Snoopy's* systems, and finally left Young alone in the CSM, closing the hatches and separating the two craft while flying on the far side of the moon. The engine burn on the LM slowed *Snoopy's* speed by just forty-eight miles per hour, but it was enough to change its orbit from a circular, sixty-nine-mile-high path, to a strongly elliptical one whose far point from the moon would stay the same, but whose near point would be only nine-and-a-half miles above the craters and mountains. Kepler's second law of planetary motion decreed that *Snoopy* would coast at its highest speed when closest, then it would slow down as it pulled away, only to speed up again as it fell in closer during the next orbit.

Cernan and Stafford had their hands busy as they stood at the controls of this new helicopter-like craft. When it swooped down to its near point with the surface, within sight of the proposed landing spot for *Apollo 11*, Gene Cernan called excitedly, "We're low, babe. Man, we're low!"

While the public usually thinks of rockets firing to "aim" a spacecraft at its desired target, aerospace engineers actually make mathematical plans with orbits alone in mind. For example, a spacecraft being fired from Earth to Mars in no way starts off pointing toward Mars. Mars would not be in that spot when the craft arrived; but more than that, the spacecraft's trajectory is also an arc and would not wind up the way it initially traveled. Instead, the craft is sent in a new orbit around the sun, one whose inward or low point (perihelion) remains Earth's orbit, but whose highest point (aphelion) is the orbit of Mars. If timed so that the craft arrives at the Martian orbit just as Mars itself gets to that spot, the two will meet in space.

Similarly, nearly all of the maneuvers and engine firings throughout the Apollo missions involved the changing of the craft's orbit to a new shape, so that it would arrive at some intended spot at a future time.

Now, once the close-in 45,000-foot-high swoops over the lunar surface were successful, the LM's ascent engine fired to change its orbit once more, this time so that its path would be boosted, to let it arrive in the same place and time as the CSM circling above.

Unfortunately, a switch had been set to the wrong position, to a setting designed to make the LM automatically seek the Command Module somewhere above. It was supposed to have been set to where it would allow the craft to remain as it was, in a stable position. Thus it was that when the crew ejected the no-longer-needed descent stage, in preparation for their ascent stage burn, *Snoopy* suddenly fired thrusters in a jerky, futile effort to aim toward an impossibly distant CM it could not detect.

"Son of a bitch!" shouted Cernan. Their LM was out of control.

But Cernan and Stafford had not spent years in training for nothing. The two test pilots regained full control of *Snoopy* in two minutes.

"I don't know what the hell that was, babe." said Cernan.

"That was something we've never seen before," added Stafford. "The thing just took off on us."

But their fifteen-second burn then went perfectly, and they were heading up to rendezvous with Young in *Charlie Brown*.

The rest of the return was uneventful. In fact, it was better than merely uneventful. The firing of their SPS engine was so precise, two midcourse correction burns were unnecessary. The crew even timed a wastewater dump at about the halfway-home mark, to let the tiny bit of thrust produced by the ejected urine tweak their speed by the minuscule additional amount desired. By the time the craft was a few hours from entering the atmosphere and a final correction scheduled, they merely used their own thrusters rather than the SPS engine, to adjust their speed by an amazing one mile per hour—about the same velocity as a toddler crawling on a floor.

When the trio stood on the deck of the aircraft carrier USS *Princeton* on May 26, the die was cast. No further rehearsals were needed, no more tests. The announcement soon came: the full landing on the moon would proceed, and the craft would leave Earth on July 16.

○

# CHAPTER 16

# Why Them?

*Oh, Moon of my delight who know'st no wane.*
—Edward Fitzgerald,
*The Rubaiyat of Omar Khayyam*

Of the four billion people on Earth at the time, why was it Neil Armstrong? Why was he destined to gain eternal fame as the first person on the moon? Why did that crew consist of Armstrong and Buzz Aldrin and Mike Collins? Did these guys have some special "pull"? Did they pay someone off?

The answer is so ordinary, so routine, it's anticlimactic.

Starting early in the decade, in the Mercury days, a routine had been established that astronauts and supervisors regarded as eminently fair. Each flight would have a primary crew and a backup crew; the latter would become the primary if the scheduled astronauts fell sick (or in two unfortunate cases, died prior to the intended flight). The backup crew would train just as hard as the primary personnel, but for their efforts would usually be rewarded with merely watching the flight on monitors from Mission Control. Their psychological compensation? Each backup crew could expect to be the primary astronauts for whatever mission lay three flights ahead. Thus the backup guys for *Apollo 8* would become the primaries for *Apollo 11*.

Neil Armstrong and Buzz Aldrin were the *Apollo 8* backups, along with Fred Haise. Haise would thus have been third crewman on the first historic landing, but fate intervened in an odd way. A bone spur had sidelined Mike Collins, who had been a primary crew member on *Apollo 8*. He therefore had to be replaced. Deke Slayton promised Collins that if he recovered from his surgery quickly, he'd be inserted into an upcoming mission. Thus it was decided that he would "bump" Haise for the third spot on *Apollo 11*.

Collins himself thought that this was fair. For starters, he had been one of the "Fourteen" selected in 1963, the third group, while Haise had been a member of the "Nineteen" recruited two years later. Thus Collins had waited two years longer than Haise for "his turn" on a moon-bound mission. Anyway, Haise would surely get his chance pretty soon. (Ironic note: Fred Haise, bumped from *Apollo 11*, did get his moon assignment—on a mission less than a year later. Unfortunately, this proved to be the near-disastrous *Apollo 13* that was forced to abort its landing attempt. Poor Haise would surely have been given yet another chance, but the final three voyages to the moon, *Apollo 18, 19,* and *20*, were canceled. Obviously, some were destined, some not.)

As for Armstrong and Aldrin, their rotation carried them into *Apollo 11*—this had been decided a full year earlier, in the summer of 1968. But even then, it was far from certain that *Apollo 11* would actually be the first to land. Any major glitches on *Apollo 9* (with the LM or with the docking or spacewalking strategies) or with *Apollo 10*'s shakedown to nine miles above the surface would mean that *11* would become yet another preparatory flight and it would be *12* that would actually land first. Moreover, in late 1968, there was even still a chance that *10* would be notched up to actually land, as some bigwigs were advocating. When the public announcement of the *Apollo 11* crew was officially made early in 1969, Mike Collins believed that his mission had only a 50-percent chance of being the first to land, according to his memoirs, *Carrying the Fire*. He thought *Apollo 10* still had a 10-percent chance, with a 40-percent chance that first honors would go to *Apollo 12* or another later flight.

Armstrong and Aldrin, it turned out, were excellent choices for the first landing. Most astronauts regarded Armstrong as the best test pilot in the bunch, a particularly high compliment since nearly all of them were experienced test pilots themselves. He was also the only civilian in that early astronaut pool, so his inclusion would have the diplomatic benefit of showing the world that while Russia only sent military officers aloft, America's first man on the moon would be "an ordinary citizen." Armstrong had another winning quality: he was not a boaster, a strutter. In fact, he so disliked publicity that the down side of this Ohioan was that he was a reluctant choice for the inevitable interviews; he frustrated the audio segments for the countless TV crews that were to follow him around before and after the mission. His answers were as short as possible, his desire for the limelight somewhere between negligible and nil.

Buzz Aldrin was another story. Extroverted and even chatty, he was widely regarded as the most intellectual, the brainiest of the astronauts. Of the early selectees, he was the only one with a doctorate, and he became widely known in NASA as "Doctor Rendezvous" for his brilliant mathematical work and amazing expertise at calculating orbits and trajectories. No one forgot that it was he who *manually* steered his *Gemini 12* craft to a perfect meeting with its invisibly distant target Agena craft after the radar computers had failed. It was also he who had indefatigably pondered, studied, practiced in tanks, and then perfected the EVA art of maneuvering outside of a spacecraft. His six-hour EVA was the one that showed why the earlier EVAs had been so difficult and arduous for the men who attempted them. He mastered it, nailed it totally like no one before him, and in doing so, significantly helped pave the way for this first landing attempt. He deserved to be here. Though a "brain," Aldrin was not a nerd; among other athletic skills he was an accomplished pole vaulter.

All told, the crew may have been just part of the rotation schedule, and it certainly can be argued that other astronauts, too, were highly qualified and eminently capable, as future Apollos readily demonstrated. Still, it was hard to top the personnel that fate had selected for the first moon landing.

Despite his unique qualifications, Aldrin barely made it. He would not have been on this crew if he had not performed so spectacularly on *Gemini 12*, and he very nearly missed the entire Gemini program. Years later, he expressed bitterness to the author at how he was initially absent from the Gemini crew assignments.

"I was an egghead from MIT, and they looked down on academics. If you weren't a Navy carrier pilot . . . well, all you have to do is look and count."

It was the last-minute death of Aldrin's back-door neighbor, Charlie Bassett, in an airplane accident, that changed everything. Now Aldrin was rotated into the final Gemini flight, and his superb performance gave him a sudden boost for an early Apollo mission. "The die rolls over and comes up seven or eleven," he mused thirty-six years later.

But even with all that, it might not have been Armstrong and Aldrin. Years later, Frank Borman insisted that *he* had been offered that first moon landing—and turned it down. The later speculation among other astronauts was that Borman's wife, Suzy, had been appalled at the risk, and successfully beseeched him not to do it. The story makes sense, especially since there had been an unusually long delay before the *Apollo 11* crew assignments came out.

As for the date of launch and arrival, no doubt NASA would have preferred July 4. Instead it was set by the moon itself. The initial priorities placed success over raw science, and NASA wanted the smoothest place possible that was near the moon's equator, above which Collins would orbit in the CSM. The Sea of Tranquillity was one of ten sites considered, and it met all the criteria. As the easternmost potential target it got the rising sun earliest. If there was a launch postponement, one of the sites to the west would still be doable without waiting a full month for the sun to come around again.

All this was dictated by lighting and, as a second concern, temperature. Anyone with a small backyard telescope knows that the fascinating lunar detail of craters and mountains shows up only along the lunar terminator, the line separating day from night. This line, which creeps along the moon's face at ten miles per hour, is where the sun is rising, where objects cast long shadows, where features have such pronounced shading that their detail becomes striking. As the sun rises higher over any particular area, the ground starts to appear stark, flat, and washed-out; the terrain heats up from the increasingly vertical sunlight, as well.

NASA's goal was to land at a flat site one day after the sun had risen there. With the sun therefore about ten degrees high and the astronauts approaching from the east with the sun at their backs, the men would see a well-shadowed landscape that would immediately reveal every boulder, hill, and obstacle for a safe touchdown. If the sun were lower still, the shadows might be too long, and dangerous boulders might be hidden in the shade of hills or raised crater rims. If the sun were higher, shadows would be too short and boulders might go undetected. If the LM came down on top of one of them and slid off sideways, there would be no way to budge it upright and leave the moon.

The Sea of Tranquillity site was just one degree above the lunar equator, and twenty-four degrees right (east) of the moon's midpoint. Anyone staring at the exact center of the moon wouldn't be looking very far from the *Apollo 11* site. When the moon is highest in the sky (due south) and the moon visualized as a clock, an hour hand positioned as in 2:30, followed for a distance just one-fourth of the way from the center toward the edge, would pinpoint *Apollo 11*.

This place would be lit up by the sun at a ten-degree angle on July 20, so the launch would have to be on the 16th. It was that simple. The moon would be a fat waxing crescent when men arrived—very much as it might have been drawn in a fairy-tale picture book.

○

# CHAPTER 17

# Stroll on the Crescent Moon

*There is something haunting in the light*
*of the moon; it has all the dispassionateness*
*of a disembodied soul, and something of its*
*inconceivable mystery.*

—Joseph Conrad, *Lord Jim*

As a media event, nothing before or since has ever matched *Apollo 11*. Historically, occurrences that changed the world were either unforeseen (such as Pearl Harbor in 1941), secret (the first A-bomb test in 1945), or appreciated only afterwards (Einstein's monumental theories of 1905 and 1915).

This was different.

Everyone knew that human history would be forever divided between the time *before* man had stepped foot on another celestial body, and the time *afterward*. Moreover, by the late sixties, manned space accomplishments had come so rapidly, everyone—absolutely everyone—expected the pace to continue. Mars would probably be reached within ten years, certainly before 1985. All of it would make the very first manned landing even the more significant.

To the delight of news organizations everywhere, this extravaganza would be precisely choreographed, with plenty of advance notice. There would be comfortable press facilities and well-thought-out camera angles for the officials, astronaut commentators, and the Saturn V launch itself. According to estimates, less than 5 percent of the world was unaware of the launch to the moon, in places like the interior of Papua New Guinea and in rural areas of communist countries, whose news policy was to stay as silent as humanly possible. For everyone else, it was no exaggeration to say that this would be the media event of the millennium.

NASA only belatedly awakened to the enormous implications, which including critical details that they themselves did not know as late as the preceding autumn. Foremost among them: Who, exactly, would be the first man to step on the moon?

By summer 1968, NASA knew who would make up the crew of *Apollo 11* (though there was still a fifty-fifty chance that a later Apollo would end up being the first to land). But even if it *was* to be *11*, Deke Slayton had still not choreographed the exact work sequence on the lunar surface.

Just a year earlier, it had been stated doctrine that only one man would venture outside the LM at a time, a policy encouraged by Sam Phillips, the NASA program director; the other crewman would tend to the ship. This had always been the way missions had been performed. You had to have some-one at the controls, minding the store.

But MSC chief Bob Gilruth strongly leaned the other way, and by late 1968 the decision had swung. The LM really didn't have to be tenanted every moment. And what if one astronaut became incapacitated on the sur-face? Shouldn't he have a companion to come to his aid? If swimmers at summer camp operate on the buddy system, shouldn't astronauts walking on an alien world? On November 1, a mere eight-and-a-half months before the scheduled launch, NASA made up its mind: Both men would be on the surface simultaneously.

Such group activity was fine, but the press demanded to know who would be first out the door. As a matter of fact, so did Neil Armstrong and Buzz Aldrin. NASA officials seemed at first not to fully appreciate that the world's press—and history books—would bestow enormous permanent honor on an *individual* rather than the entire collective American effort.

In truth, there were dozens of astronauts, hundreds of supervisors and planners, a thousand engineers, and a half-million workers who had made this happen. The individual who first stepped down the ladder was merely the guy whose turn it was to sit in the left seat. Why did *he* deserve all the glory? NASA had a point. If one were to try to find "most important" indi-viduals that had made Apollo a reality, it would be impossible to ignore flight controller chief Chris Kraft; Apollo program director Sam Phillips; as-tronaut chief Deke Slayton, whose own wings had been so unfairly clipped; Bob Seamons, who was essentially NASA's "general manager"; Marshall Space flight director Wehrner von Braun, whose rockets had started it all; NASA administrator Tom Paine; MSC (Manned Spacecraft Center) director Bob Gilruth; KSC (Kennedy Space Center) director Kurt Debus; Langley Research Director Edgar Cortright; KSC flight director Rocco Petrone; Lewis Research

Center director Abe Silverman; Electronic Research Center Director Jim Elms; Mission Operations Control director Gene Krantz; a dozen other directors who worked on Apollo primarily through such advanced academic centers as MIT; and hundreds of brilliant and innovative designers, engineers, and problem solvers who never did get the credit they richly deserved for making the whole thing possible.

And for all that, the press was going to glorify a single test pilot?

Exactly. By early spring of 1969, it was apparent that "the first man on the moon" was going to be an incredibly big deal. So, again, who would it be?

The LM crew were intelligent men whose sense of history was in no way absent. Despite being fundamentally shy, Armstrong was a doer, an accomplisher, and more than bright enough to realize the awesome implications. Within their insular community, he and his competitor made no secret of the fact that each wouldn't mind being the first out the door.

If past precedent had been followed, it would have been Buzz Aldrin. He was the Lunar Module pilot, this was "his ship" that they'd be piloting to the surface. During all earlier discussions of the matter, it was assumed that the LM pilot would be first down the ladder. On the other hand, Armstrong was commander of the mission, which gave him the highest rank. And as every military person in the world knows, rank hath its privilege. Moreover, the hatch was closest to Armstrong. If Aldrin were to be first out, he'd have to climb over his commander.

The thing was, in every Gemini and Apollo flight to date, the commander would always stay with his ship while a more junior officer exited from the module to perform any extravehicular activity. The commander simply never left the ship first, it was an established custom. But again, the commander *was* the commander, and he *could* attempt to exercise his prerogative if he so desired.

One could see it both ways. In the end, it was top MSC play-callers, including Deke Slayton, who would create the work-chart scheduling.

The press, for their part, assumed that past routine would be followed. When they learned the crew member identities in 1968, they widely speculated and reported that Aldrin would be the first man on the moon. The media responded with no small amazement when, in early April 1969, a mere fifteen weeks before launch, NASA announced that it would be Neil Armstrong.

Buzz Aldrin tried to retain his composure, but everyone who knew him saw that he was seriously nonplussed. This was *his* job! He had practically *tasted* the "first on the moon" glory. The newspapers had even reported it. As Mike Collins later recalled, "Some of the early check lists were written to show [an Aldrin] first exit, but Neil ignored these and exercised his commander's prerogative to crawl out first [when rehearsing the mission]. This had been decided in April, and Buzz's attitude took a noticeable turn in the direction of gloom and introspection shortly thereafter."

Collins went on to say in *Carrying the Fire*, "Once, he [Aldrin] tentatively approached me about the injustice of the situation, but I quickly turned him off. I had enough problems without getting in the middle of *that* one."

But to the cameras and the reporters, Aldrin forever kept his cool and said it didn't matter. He was a professional. Anyway, what could he do?

# The Day

No rational person lingers outdoors in Florida in July. But nearly a million people did exactly that, emerging from their tents and campers, coffee in hand, to stare across the causeway at the giant rocket, Mission number AS-506, Project G, known to everyone outside NASA as *Apollo 11*. First blushes of twilight found the day nearly windless and already swelteringly humid, and by launch time in midmorning the thick unpleasant air would rise to only a hair under ninety degrees.

The three astronauts were awakened at 4:15 a.m. A black moonless sky hung outside their windows as they sleepily breakfasted on steak, eggs, juice, and coffee before getting suited up to begin breathing 100-percent oxygen—necessary to purge their bodies of nitrogen that could otherwise bubble painfully out of their blood at the low one-third sea-level pressure that their ship would maintain once in space.

So now, already isolated from Earth's atmosphere even though their departure was still three hours away, their van left their facility at 6:27 for the slow five-mile drive to launchpad 39A, with Venus, the morning star, still brilliant in the brightening eastern twilight. By the time they arrived at 6:51, the sun was just above the horizon. The ancient Greeks believed that every successful journey should begin in the morning, and the trio, as if following the advice, stepped from the transfer van into a newly risen dawn. They gaped up at the thirty-six-story behemoth that would take them all the way to the moon.

In the past, this area around the rocket, with its equally tall gantry, had invariably swarmed with workers and technicians, but now it was virtually deserted. To the astronauts' eyes the scene looked strange, even eerie. Explanation: Only the barest skeletal crew was allowed in the vicinity of this monster, now that it was laden with explosive fuel.

The men rode the elevator, and Armstrong stepped into the Command Module first, Aldrin last. The countdown was old news for the trio, all of whom had been in space before, though none aboard a Saturn. When the countdown reached zero at 9:32 a.m., the noise became deafening and the vibrations were substantial; the men were continuously jerked sideways against their harnesses as their bodies were increasingly pushed ever-deeper into their seats by the rising g-forces.

As always for the Saturn V, the first stage's five F-1 engines, each burning 3,000 pounds of fuel per second, lasted for just two minutes, forty seconds. At that point, at an altitude of forty miles and a speed of 6,000 miles per hour, the second stage kicked in with its five J-2 engines. This time the dangerous, scary, and usually erratic second stage flew with an odd smoothness. In all this upward acceleration the g-forces never exceeded 4.5, which was not unpleasant for the experienced crew. (They would endure double this when returning to Earth.)

The single J-2 engine of the third stage then fired. A mere eleven minutes and forty-two seconds after launch, the men were in Earth orbit, suddenly floating, suspended in their harnesses upside down with their heads pointed toward Earth.

Their next two-and-a-half hours circling Earth were busy checking all systems before the all-important TUI, the further firing of the third stage to place them into trans-lunar injection. Everything did indeed check out, approval was given, and a five-minute further burn pushed the men with a one-g Earthlike sensation whose meaning was: You're outta here!

When the fire stopped they were ascending rapidly toward the east at a whopping 24,000 miles an hour, enough to escape from their home planet. Coasting now for the next three days, this was the fastest speed they would achieve on the outbound leg because Earth would tug at them the whole time, eventually slowing them to a paltry 2,039 miles per hour before the moon took over and sped them up once more.

*Apollo 11* reached this equigravitational point sixty-one hours, forty minutes into the mission, when the craft was 88 percent of the way to the moon. Now the pace of events began to quicken. They had long ago discarded their spent third stage, had flipped around to pull the LM from its storage "garage," and fired their own SPS engine. The next decisive milestone was to occur in the moon's vicinity just minutes before the men would vanish behind its western edge (when they would for the first time be cut off from all radio contact with Mission Control); it was the go-ahead for the long six-minute SPS burn that would slow them down by 2,000 miles per hour and insert them into a lunar orbit.

It would have to happen on the moon's far side. If the engine failed to fire, *Apollo 11* would reappear twenty-three minutes later, on a looping trajectory that would fortunately throw them roughly back toward Earth. If all went well, their reappearance would be delayed by an extra ten minutes, and they would be firmly captured by the moon.

Their voices came through at the expected moment, and after making two elliptical orbits around the moon, they fired their engine once more to circularize their path. They were in orbit around Earth's nearest neighbor.

But—bizarre at it seems—they weren't alone. Unbeknownst to the world at large, the Soviets had tried one final, desperate attempt to steal some of the moon-landing thunder. Just two days before *Apollo 11* had left Earth, the *Luna 15* spacecraft had been launched—for the moon. And it was now in lunar orbit with the American astronauts. This three-ton Russian craft had the capability of soft-landing on the moon and actually analyzing the soil and rocks. The Russians had intended to grab some of the spotlight by announcing: See, we can go there too and accomplish the same scientific exploration, but without the risk to human life.

In a rare moment of cooperation, the Russian and U.S. mission control teams had shared last-minute information about each craft's intended orbit around the moon, in hopes of avoiding the super-slim chance of a collision. The three American astronauts—alone among the millions of people watching their TV sets back home—had been informed of this secret massive Soviet craft now sharing their lunar orbit. They weren't worried. "Space is large," one of them shrugged, years later. (As an epilogue, *Luna 15* ended up crashing onto the surface and thus performed no analyses whatsoever. In typical Soviet fashion, the existence of this attempt was simply never announced.)

Now Aldrin and Armstrong crawled through the tunnel into the LM, checked out all its systems, and separated the two craft.

This left Michael Collins alone in the CSM. He would spend two days as the loneliest man in the universe, especially when he would pass behind the moon out of radio contact with every other human being. No, more than that: Out of the four billion people then alive, he'd be the only one unable to *see* Earth. And, also unlike everyone back home who had a TV on, he'd be unable to watch the goings-on of his compatriots down on the lunar surface.

He was also keenly aware that if something went terribly wrong on the surface, if, for example, the LM's single ascent engine failed to fire, or if, more likely, the landing knocked the LM over on its side, his associates would soon die on the moon and he would return to Earth alone. This indeed was the main reason that *he*, and not one of the others, had been trained as the CSM pilot. He could easily bring his ship back all by himself.

History's recollections about *Apollo 11*, however, were always destined to give short shrift to Collins, and focus instead on the two others who now separated their LM from his CSM and rotated their odd insect-looking machine so that it was aiming engine forward. In the newly resurrected tradition of giving spacecraft names instead of number-and-letter call signs, Collins's CSM was *Columbia*, and the LM was called *Eagle*. Indeed, as soon as Armstrong and Aldrin had separated their craft and started its independent flight, they radioed to Houston that "the *Eagle* has wings!"

It should never be imagined that the men had merely climbed in and hit a few switches. On a mere two-hours' sleep, and with a grueling twenty-two-hour period of wakefulness ahead, the astronauts had spent several lunar orbits and hours of intense activity working with Houston checking innumerable systems. Antennae were aligned and collimated, navigation systems inspected and tested, electrical equipment checked and cross-checked, numbers and figures read back and forth to Earth, propulsion and environmental systems activated and analyzed. The unrelenting pace, the sheer number-heavy workload could have made lesser men mad.

But now they were separated, and, in radio silence on the far side of the moon, they ignited the LM descent engine. This firing would reduce their forward speed by just fifty miles per hour, but it would change their path from the present circular track to an elliptical one whose high point, or apocynthion, was sixty-nine miles above the surface and whose low point, or pericynthion, on the moon's opposite side would be just ten miles. When the LM reached that low point, it would again fire its descent engine to further slow it to a decreasing speed and a gentle touchdown. That was the part of the plan that had never been performed or rehearsed before, ever.

Up to this point, nothing terrible was realistically expected to happen. If the descent engine now refused to fire again, the LM would simply loop back up from that low ten-mile pericynthion back to the sixty-nine-mile-high apocynthion, the Command Module's altitude. The CSM could then maneuver to link back up with the faulty LM. No, if things were to get hairy, it would probably be during the final break-free from orbit when the craft was descending the final few miles to the surface.

The world was watching, but the greatest apprehension naturally hovered on the shoulders of the people who had made the whole thing possible. They crowded behind glass at the back of Mission Control, watching the 145 controllers who themselves were nervous and perspiration-covered. This was it.

The powered descent, or PDI, went uneventfully at first. The long rocket firing, with the men upside down, would slowly transition to a nearly upright descent, as they went down from 50,000 feet and a speed of 2,000 miles per hour. In the interim, the LM would fly 300 miles along the surface, from night into day, right to the intended touchdown spot on the edge of the Sea of Tranquillity.

But it was not destined to be smooth and uneventful.

First, communications became terrible. Houston could not hear the LM, and the men could not hear Houston. Everything had to be relayed through Collins, orbiting above. Then, telemetry data from the LM kept freezing up on the screens back at Mission Control. Gene Krantz was in charge as flight director, watching everything, and he decided that even if his men were receiving old data, he'd let the pair continue downward so long as none of the data was bad. So long as everything stayed within normal ranges they could continue, even though he wasn't receiving the latest.

It must be remembered that this was very much a group effort. Armstrong and Aldrin could not possibly monitor the health of all their systems. This vital function was being accomplished on the ground, with relevant information from multiple sources distilled down and relayed to Armstrong and Aldrin by a single person, the CapCom—who was Charlie Duke. Let's put ourselves in on the action:

Duke is hearing the relevant fragments of information from the various controllers and telling the men only what is necessary for them to digest, since they're up to their neck in alligators simply doing their job of landing.

Nobody yet realizes it (they would in a few minutes) but when the men separated the LM from the CSM, there was a little air in the connecting segment, which, like a champagne cork popping, blew the two ships apart with a bit more speed than was expected. Now this extra velocity is coming back to haunt them, because they are heading toward a point at the extreme upper end of their landing "footprint"—the elliptical zone in which they had planned to touch down. As a matter of fact, they'll be inadvertently heading for the very "toes" in that footprint, the edge, in which will lurk a boulder field on which they cannot possibly land.

Now an alarm sounds, the 1202 alarm, which means that the computer is overloaded. This alarm signifies that the computer doesn't have enough time to do what it's being asked to do, so it's going to start prioritizing— do the most important stuff, skip the rest. Should Mission Control order an immediate abort, and a firing of the LM's ascent engine? The rules were clear: if it happens again, they would have to abandon the descent. Instead Mission Control comes up with a quick fix and takes over some of the computer's functions so that it is able (everyone hopes) to handle its workload.

Communications have now thankfully returned, splendidly. Everyone can hear everyone else, though the telemetry data still gets erratic. They keep going. At six-and-a-half minutes into the descent burn, about halfway, it's time to bring the engines down to 55-percent power. The astronauts are now less than five miles above the surface (23,000 feet) and have slowed to 1,000 miles per hour—still breakneck speed. If all goes according to plan they are less than three minutes from touchdown.

At his console, Bob Carlton monitors fuel consumption, and now he says, "low level." This simply amazes Flight Director Krantz and everyone else. In the endless simulations this point had not been reached until the men were already on the surface. Carlton has an assistant, a controller named Bob Nance, who has been staring at a recorder monitoring every moment of Armstrong's throttle control. Nance keeps a timer on how long the engine is firing, and at exactly what setting, and is so amazingly skilled at calculating fuel quantity that in the simulations he would call out when the fuel would be gone and always hit it to within ten seconds. He now grasps that they're already low. From this point on, it is "running out of gas" that is the most serious issue, and Krantz worries increasingly that they're going to have to abort the landing.

But they can't safely do so without having a few seconds of usable fuel in the descent engine, because they'd first have to throttle it up and gain a positive upward rate of climb before they could safely separate from the landing platform and fire the ascent engine. They cannot simply wait to run out and begin dropping down. Then it would be too late.

At eight-and-a-half minutes into the burn, the spacecraft switches to a different computer program: they've finished their "braking" segment and are starting their "approach" phase. At this moment, only 7,000 feet up, the men are moving forward at 300 miles per hour and downward at 87 miles per hour. From this point on, with the craft nearly upright and tilted back only slightly at thirty degrees from vertical, the crew could take over manual control and help steer the LM to a smooth touchdown spot.

Mission controllers, monitoring all the LM's systems, still make a series of final go/no-go decisions based on their telemetry. Gene Krantz takes a final poll and all systems monitors, scanning incoming LM data on their screens, shout back "Go!" in turn.

The tense but hopeful control center then gets nerve-wracking news: another loud alarm. This time it's a 1201, the code for another computer overload. The onboard brain is perilously close to having so much to do that it will soon start dumping valuable data. Happily, the LM is now on the verge to switching to the next stage of descent, the final "landing phase" when that program will no longer be needed. The men get a go-ahead to ignore the alarm.

Armstrong switches to computer program P66—the final landing guidance. He's just 400 feet up now, forty stories, his LM angled back just eleven degrees, very nearly vertical. If he lets the computer automatically do all the work using its onboard radar, it will continue descending while moving forward only another half mile, then, hovering, slowly descend to a touchdown. But Armstrong thinks he can now clearly see where that eventual touchdown spot is located, and he doesn't like it at all. An enormous boulder field fills the location; landing on any of them would topple the LM. Just as bad, a substantial crater with dangerously sloping walls lies immediately beyond the boulders. (Later analysis showed that Armstrong misjudged the LM's automated trajectory. They would *not* have landed in the boulder field, nor the crater, but would have gone to the right of both of them and settled in the area of three smaller craters. No matter, *that* spot was nearly as bad.)

Armstrong assumes manual command, grabbing the controls to take over from the computer. He scoots the LM forward, not just a little, but a considerable distance. He feels he really doesn't have much choice, but he knows he doesn't have much fuel either. (Much later, back on Earth, reflecting on this whole process, he apologizes for what he perceives as a poor landing job. The rest of the world knows he's being way too hard on himself.)

He now maneuvers the craft sideways at ten to fifteen feet per second, far faster than had ever been practiced in the simulators. He is trying to find and reach a smooth place while he still has fuel.

"Sixty seconds!"

It's Carlton's voice, calling out the amount of remaining fuel burn time.

Armstrong still isn't even close to the surface, but he keeps going. Aldrin standing next to him neither agrees nor disagrees, for he hasn't had a moment to look out the window. He's keeping his head down reading the instruments and calling out to Armstrong vital information about speed and remaining distance to the ground.

Gene Krantz mutters, "Boy, he's going to really have to let the bottom out of this pretty soon."

CapCom Charlie Duke is still calling out information to Armstrong and Aldrin, but then Deke Slayton, sitting next to him, gives him a shot in the ribs.

"Shut up now. Just let them land."

Duke simply nods. At an altitude of 120 feet, the descent engine is throwing up enormous amounts of the fine surface powder and it starts blocking out the lunar terrain below them. This material spreads radially away from the spot directly beneath them, and obscures almost everything. Armstrong is now descending blind. At a height of 80 feet, the density of material increases substantially.

They're still eight stories up. If the engine cuts out now, they'd fall the remaining distance, wrecking the ship and stranding them, even if they survived the plummet.

Armstrong has now brought the LM to a near-hover, moving sideways only a little as he stares out at the thick impenetrable powder being thrown below them. From brief glimpses through the dust, he thinks he can see no large obstructions below. The craft is still ten or twenty feet in the air.

"Thirty seconds!"

It's Carlton again, the only voice Krantz now allows. At this point it will be fuel alone that could cause an abort, nothing else. Why confuse the men with anything else?

This was cutting it way too close. The fuel gauge had an inaccuracy of 2 percent, which means the engine really could cut out any moment now. But Carlton and Nance know this fuel business well. They think the men have between fifteen and thirty seconds. Still, jeez, they *had* better put it down or forget the whole thing. No more time for smooth spots. It's now or never.

Buzz Aldrin still reads the instruments, trying to help Armstrong during these final few feet to the ground:

"Forward. Drifting right! Contact light!"

"Contact light" is the best thing everyone can possibly hear. Three of the four legs have thin probes that will activate a bright blue light on the panel when they touch something, anything.

"ACA out of detent," says Armstrong. He shuts down the engine at the same instant that Carlton says, "fifteen seconds."

The next voice comes from Aldrin:

"Okay, engine stop."

Thus, the first word spoken from the lunar surface comes from Buzz Aldrin. It is: *okay*.

Cheering erupts at the control center. Everyone knows what "engine stop" signifies, because the astronauts are supposed to just let the ship drop down the remaining few feet as soon as they see that blue contact light. There's no need to wait for any special speeches from the guys. They are on the moon.

The astronauts continue their turn-it-all-off checklist, as Houston hears them say, "Modes control both auto. Descent engine command override off. Engine arm off. 413 is in."

CapCom Charlie Duke can wait no longer:

"We copy you down, *Eagle* . . ."

And Armstrong, reasserting his command role, states the obvious:

"Houston, Tranquillity Base here. The *Eagle* has landed."

○

# CHAPTER 18

# Adventures on Tranquillity Base

*Meet Me By Moonlight Alone.*

—Joseph Wade

Cheering broke out throughout Mission Control. The men were down and safe. They had done it. And "Tranquillity Base" had a wonderful ring to it. Nobody had ever thought of using such a term in the simulations, or describing the landing spot as any sort of base.

But not everyone could relax. What if the surface collapsed under them? What if some great danger lay seconds away? What if the remaining propellant was escaping due to some landing-produced jolt, and it was pooling below the LM, ready to explode? Obviously they must first quickly check out all systems and then make an immediate stay-or-leave decision. Two possible escape windows existed, both of which could get the *Eagle* up to an emergency rendezvous with *Columbia*. One was immediate, the next came in nine minutes.

But all systems seemed okay, and they were allowed to stay. When the nine minutes had come and gone, the next launch opportunity wouldn't arrive for two hours.

After everything was carefully monitored and systems were checked, tank gauges inspected for the possibility of leak or rupture, and the computer set for the program it would immediately need if they were to leave quickly, everyone started to relax. For better or worse, they were now on the moon for at least two hours. There was no longer any need for speed or tension.

They had landed four miles southwest of their intended touchdown point, but it hardly mattered. Aldrin delivered a description of what he saw out the window, and then a prayer. Houston offered the men a chance to sleep, but the twosome was far too keyed up to even think about napping. The astronauts instead asked if they could begin their spacewalk, and NASA approved it. There had been much back-and-forth about this very issue for weeks before the flight. If the original schedule were to be followed, they'd be walking when nearly everyone in the United States, and most in Europe, would be asleep, in the wee hours of the night. By getting the go-ahead now, their moonwalk would happen in prime time.

It took nearly three hours, which seemed interminable, for the men to get their cumbersome suits and helmets on, and check out the radios and oxygen and all the rest. In fact, it was fully six hours after they landed, or 109 hours, 2 minutes after launch, that Armstrong finally opened the little square hatch, measuring a mere two feet, eight inches on a side.

It was late at night on the twentieth, or, for Europeans, early in the morning on the twenty-first, when Neil Armstrong emerged, backward, and started down the ladder. It wasn't easy for him to see how high up the final, lowest rung was: the ground beneath the ladder was in shadow. On the moon all shadows are pitch black, unlike on Earth where air always scatters light and the blue sky fills in the areas out of direct sunlight. (Artists have always realized that the shadows of trees on white snowy surfaces look blue, not black, for this reason.)

So Armstrong, heading down feet first, could not fully see where he was stepping. Instead, he tentatively hung a leg downward into a black empty space below the final rung, like a bather testing the water temperature. Viewers at home didn't know it, but the landing plan had called for shutting down the engines as soon as the "contact light" had come on, and letting the LM drop the final three or four feet. This would have accordioned the deliberately compressible honeycombed legs. But Armstrong had instead been so gentle in his descent that the engines were still firing when the LM was fully on the surface! As a result, the legs were in no way compressed, and the ladder was therefore higher up than expected, with a substantial three-foot drop beneath the final rung. Armstrong now saw all this, but if he was in any way concerned he didn't show it.

"I'm at the foot of the ladder. The LM footpads are only depressed in the surface about one or two inches. Although the surface appears to be very, very fine grained as you get close to it. It's almost like a powder. Now and then, it's very fine."

The world held its breath. Would he *ever* step off, and then say whatever it was that the press had speculated about for months?

It had been a topic hotly discussed: The first human words from the moon. What would they be? Many hoped that Armstrong would avoid any parochial or patriotic sentiments, that his speech would resonate with some universal theme. It was widely believed that the words would be remembered for all time.

At this point, few at home realized that "Okay, engine stop," spoken by Aldrin six hours earlier, had *really* been the first words from the moon. But even if they did, the ones that somehow were deemed to be "official" would be the words uttered from *outside* the spacecraft, not from within its walls.

NASA had insisted that Armstrong was at liberty to say whatever he wished; it was one more manifestation of America being a free country. When his own crewmates brought up the topic of what he intended to say, during the three days that they were en route to the moon, Armstrong had been evasive and ambiguous. He wasn't letting on that he had even decided yet. And maybe he hadn't.

Of course, many suspected that *someone* from NASA's PR department must have secretly met with Armstrong sometime before the flight, and either made discreet suggestions, or else asked to hear what the Ohioan had in mind. After all, this was big. Would NASA really leave it entirely to chance? Armstrong was a test pilot, not an orator.

The television camera, marvelous in that it was showing anything at all, but criminal in that the black-and-white images were crude and ghostlike, especially when compared with the vastly superior quality of what was televised during later landings, now showed Armstrong reaching his leg widely out and down, and then hopping down

"That's one small step for man, one giant leap for mankind."

Everyone cheered. Announcers repeated it over and over. Translators quickly rendered it comprehensible in every language. Of course, there was one little problem: it made no sense.

At least that first clause required interpretation. The final clause: "one giant leap for mankind" was fine; its meaning was obvious. But what did that initial "man" signify in *"That's one small step for man"*? Did "man" mean *mankind*, as it usually does when stated without an article? Or had Armstrong meant *himself*, in which case he should have said, "a man."

Had he blown his lines, like someone's over-rehearsed speech to the boss? Later, when asked about this, Armstrong said that he *had* said "a man," and that it must have been lost in transmission. Nobody believed him. The tapes were crystal clear, with no other "dropouts" or other missing segments or syllables anywhere else. He had clearly said, "One small step for man." Decades later, in 2006, an audio expert, analyzing the original recordings, claimed that he had found evidence for the infamous dropped "a" and that Armstrong had therefore uttered it after all.

No matter. NASA officially revised the speech in the 1970s, airbrushing the article in as if it was a Stalinist era photo, and today some textbooks quote Neil as saying "step for a man" while others quote "step for man." And so the little speech was either eleven or twelve words long depending on the version one prefers.

Nineteen minutes later, it was Buzz Aldrin's turn to back out the hatch. Armstrong had his camera out and filmed the whole thing.

Aldrin: "Now I want to . . . partially close the hatch. Making sure not to lock it on the way out."

Armstrong: "A good thought."

Being locked out of the spacecraft? Was this even possible? Was this some kind of comedy routine?

Armstrong offered more helpful advice to Aldrin: "You've got three more steps and then a long one."

Aldrin hopped off, and the long one proved a bit too long for him. As he made contact, he felt the urine bag in his left boot break open, and felt warm urine flowing around his foot. He decided not to say anything; he knew he had an "open mike" and that the world was listening.

"Magnificent desolation," said Aldrin instead, quickly demonstrating his forte of being more articulate than his commander. He then joined Armstrong, taking "one small squish" and many others as he walked the surface helping to set out various arrays of experiments. They were all part of the EASEP, the Early Apollo Scientific Experiment Package. They also hurriedly gathered some soil samples just in case a later emergency necessitated a quick departure. Whatever else might happen, they weren't going to leave empty-handed.

They tried different ways of walking and hopping in the one-sixth gravity, and wound up walking like back at home, one foot in front of the other. Starting with the very next mission, other astronauts came to a different conclusion, and all ten future visitors found it easier to hop like kangaroos.

Neil Armstrong, so competent and professional, now allowed himself his sole moment of joy. Swinging his arms over his head and looking around the moon's stark gray terrain, he said, "Isn't this fun!"

It was. Later astronauts universally reported that the one-sixth gravity of the moon's surface was far more enjoyable than being weightless.

Then the men heard from Houston, asking them to pose in front of the TV camera while they accepted a phone call from the president.

Poor Richard Nixon. This whole moon thing was a nightmare. It had been his archrival's idea, and he had to stand by gamely listening to announcers forever alluding to Apollo as meeting John F. Kennedy's visionary goal. Nixon hated Kennedy, the man who had beaten him in 1960. Kennedy had been charming and handsome, Nixon gawky and homely; it had been the suave patrician versus the used-car salesman with the five o'clock shadow. Kennedy, now dead from an assassin's bullet for six years, had been popularly elevated to the stature of saint and martyr. For his part, Nixon was embroiled in an increasingly unpopular war, and this moon landing should have offered both distraction and glory to the president of the nation that had pulled it off. Except—it was Kennedy's name that was still attached to it!

Yes, Nixon disliked the whole Apollo business, and indeed canceled all the later flights as soon as he could pull it off. But, now, always the consummate politician, he wasn't about to let the limelight shine anywhere so brilliantly without stepping into it himself—even it happened to be off-planet. While Armstrong and Aldrin stopped their work to essentially stand motionless at attention, breathing their canned oxygen, Nixon rambled on:

"Neil and Buzz, I am talking to you from the Oval Office of the White House. . . ."

It was the most expensive long-distance call ever made, but Nixon continued for 160 words while the men stood in the glaring sunlight, Aldrin in his wet urine-soaked boot.

At the White House's insistence, Nixon had also affixed his name *and* his signature to plaques attached to the LM's legs, that would be left on the moon at each of the Apollo landing sites. His was the only repeating name on all of the plaques, a total of twelve "Richard Nixon" signatures at the various lunar locales. They should remain legible in the moon's airless environment for many millions of years. Long after the human race is either gone or evolved to something unrecognizable from today's version, aliens landing on the moon in the far future and finding these plaques might well wonder who this "Richard Nixon" was, what unique accomplishment or status justified having his recurring name chosen as the only one to survive the human race.

Phone call over, Armstrong and Aldrin continued to set up their experiments. This included moonquake recorders with a seismometer so sensitive, it later detected the ground trembling from the impact of the men's urine bags, when Aldrin tossed them from the LM prior to liftoff. (Another "first": *Apollo 11* marked the first time that pee had been detected by seismometers.)

They hammered a tube into the surface to take a sample, but found it extremely difficult to sink anything more than five inches. On Earth, soil always has air spaces. On the moon, there is no air even in the ground, so materials become extremely dense just a few inches down.

The men took their special American flag with a deliberate fakely furled faux-windblown appearance and set it up next to the LM. Here, too, they found it very hard to push the staff into the lunar surface. Although they managed, it promptly and permanently fell over when they later fired their ascent engine to leave.

## Moon Landings a Hoax?

Fueled by pseudo TV documentaries and endless Web gossip, many believe that we never went to the moon, that it was all an elaborate publicity stunt, shot on a sound stage somewhere. Among the reasons usually cited are the planted flag seemingly blowing in the wind ("There is no wind on the moon!" exult the conspiracy believers) and the absence of stars in the background black sky.

Contradicting such silliness are the hundreds of thousands of people who actually labored on the Apollo program, the obvious Saturn rockets blasting off and going *somewhere,* and the silence or acquiescence of our enemy the Soviets who would have loved to discredit the project if it were remotely possible. Moreover, the corner cubes left behind on the lunar surface by several Apollo teams remain even today—and reflect laser pulses back to all universities and labs that shine light on those locations.

As for the specific no-wind arguments, these are easily dismissed. *Of course* NASA knew that the moon is windless—this is common knowledge to every sixth-grader: If a hoax had been planned, would they really be dumb enough to show flags waving in the breeze? Instead, these were special plastic banners designed to give the appearance of motion, rather than having our national symbol hang limp. Regarding stars, these couldn't show up because the cameras' apertures were set to show images in bright sunlight, thus underexposing the stars to oblivion.

Finally it might be noted that the actual task of pulling off such a stunt, including the hiring of actors, gaining the dishonest acquiescence of the many astronauts, keeping the silence of the camera crews, set designers, directors, and so on, and their wives, girlfriends, and boyfriends, and maintaining this secrecy for decades without the slightest leak—would have been more difficult and implausible than actually going to the moon in the first place.

The Apollo team also placed on the lunar surface the first corner reflector, made of fused silica. Like highway signs that reflect headlight illumination directly back toward whatever car produced the light, these would precisely return any laser back toward whomever beamed it at the moon. In coming months, years, and decades, universities and other research facilities would send laser pulses to these various Apollo landing sites and, by precisely timing the return flash some three seconds later, could measure the moon's distance to the nearest inch.

The men also scooped up more lunar soil. After those initial grabs they could be more leisurely, and now took careful samples in well-chosen areas, a total of forty-eight-and-a-half pounds.

The astronauts could not of course know what the lunar rocks and soil would contain, but later analysis showed a wholly unknown mineral rich in oxygen that also included iron, magnesium, and titanium. In the *Apollo 11* astronauts' honor it was named Armalcolite, after Armstrong, Aldrin, and Collins.

After a mere two hours, twenty minutes outside, both men were back in the LM, their hatch sealed, the interior pressurized with oxygen. In the cramped space, they could barely find space to stretch out, and slept only fitfully after Armstrong elevated his legs in a makeshift sling.

When they were awakened to prepare to leave, they first put on their suits, depressurized the LM, and tossed out everything that was no longer necessary, to save weight.

People often ask about space litter, the garbage left behind on the moon. Considering the moon's surface area of over thirty million square miles, the materials left behind were minuscule. If one didn't know exactly where to look, they'd be almost impossible to find. But, for the sake of thoroughness, here is an inventory of what the astronauts left:

1. Two life-support backpacks.
2. The descent stage, with the famous plaque attached to one of its legs.
3. A gold olive branch, symbolizing peace.
4. A large bag of unneeded gear such as armrests, brackets, cameras, and tools. Also, those filled urine bags.
5. A shoulder patch from the tragic *Apollo 1* mission, honoring Gus Grissom, Roger Chaffee, and Ed White.
6. A small ping-pong ball–sized disk containing micro-etched messages of goodwill from the leaders of seventy-three countries.
7. Medals honoring Russians Yuri Gagarin and Vladimir Komarov, who had each died during their country's space program.
8. The American flag and its staff, now lying on the ground.
9. The mast of the solar wind experiment (a thin sheet of material that had been placed on it and exposed to the sun for more than an hour had been removed for transport back to Earth).
10. The corner cubes of the laser reflector unit.
11. The seismic experiment.

After twenty-one hours, thirty-six minutes, thirteen seconds on the moon, the ascent engine fired. There had been one worrisome aspect to this. Early after the landing, Aldrin had swung around and his backpack hit the engine arming switch and broke it off. This wasn't some small item: it was the circuit breaker that electrically activates their ascent engine. At any rate, before liftoff, when they came to the part of the checklist where they push in the breaker, they reached in and depressed it with the tip of a pen, and fortunately it stayed "caught," so that was that. Aldrin saved the broken breaker plastic as a souvenir, and brought it back to Earth where he still stores it in a safe-deposit vault as a keepsake. At 1:54 p.m. on July 21, 1969, the five-ton *Eagle* lifted off from the moon. Because of the weaker gravity, the force that was required was only the same as boosting a small car from Earth's surface.

The engine *had* to fire, or else they'd never get away and never meet Collins in the orbiting CSM. When it did, it seemed as if the last hurdle had been crossed. They left their descent stage platform on the surface and headed up. At 160 feet their computer took over and slowly pitched their craft on its side for the angled approach to orbit. Now, without all the dust that was kicked up during their descent, the men could admire the view. They quickly recognized craters they'd studied:

"There's Sabine off to the right—there's Ritter out there. There it is right there! Man, that's impressive!"

The engine burn lasted seven minutes, twenty seconds and smoothly boosted them to a speed of 4,138 miles an hour.

When the engine stopped, *Eagle* was in a highly elliptical orbit whose low point, arrived at this moment, was ten-and-a-half miles high. Following this looping, coasting trajectory while they headed around the moon, the LM ascended toward the orbit's apocynthion of fifty-two miles. When it reached that point, their engine would again fire to place them in a nearly circular fifty-two-mile orbit. From here, orbiting not too far below *Columbia*, further burns would cause them to close their distances.

Computers as well as manual controls on the two craft were performing this work, with Houston mostly standing by and eavesdropping. Since all three astronauts were Gemini veterans of orbital-rendezvous procedures, *Apollo 11* was in capable hands. Indeed, they had Doctor Rendezvous himself aboard.

When the two craft finally met there was a new procedure, because the LM was filled with loose moon dust that could float through the tunnel into the CSM and clog things up. Armstrong and Aldrin carefully vacuumed everything they could, while Collins overpressurized the CSM so that there would be a flow of gases toward the LM rather than from it. Still, Collins got a whiff of the powder from the moon, and thought that it smelled awful.

All this happened while the linked spacecraft were on the far side of the moon. When they appeared, Mission Control was ecstatic to hear: "Houston this is *Columbia*, reading you loud and clear. We're all three back inside, the hatch is installed. We're running a pressure leak check. Everything's going well."

The Houston controller, a little surprised but delighted at how fast things were going, radioed back: "Roger *Eagle*, correction, *Columbia*. We copy. You guys are speedy."

The crew, an hour and a half ahead of schedule, then fired small explosives that separated *Eagle*, and sent it falling downward where it would crash into the lunar surface.

Finally, 135 hours after their launch from Earth, while again on the far side of the moon, the CSM's powerful engine fired for two minutes, twenty-eight seconds to raise their speed by 2,237 miles per hour—and they were now on a three-day trajectory toward Earth.

After checks of navigation and other systems, the tired crew—they had been essentially up for more than forty hours, since they slept only fitfully in the LM on the moon—were given a nine-and-a-half-hour rest period. Houston posted an audio "Do Not Disturb" sign on the craft.

Aldrin, Armstrong, and Collins had earned it.

○

# CHAPTER 19

# Struck By Lightning

*Oh Moon of Alabama*
*We now must say goodbye.*
—Bertolt Brecht, "Alabama Song"

ost people can easily name the first person who sailed around the world. But as for the second man who accomplished the feat—can anyone recall who that was? That sort of limited name recognition may have to do with the mind's finite storage capacity, but it's probably simpler: being first is everything. So it is that the ten people who walked the moon after Armstrong and Aldrin are, sadly, all but forgotten.

It's a shame. The subsequent missions produced a much greater level of science, generated adventures just as exciting and even more surprising than the first, happened in places that were far more dramatic and interesting, and incurred every bit as much risk as the groundbreaking odyssey to the Sea of Tranquillity.

The decline of the later Apollos in the public mind was accompanied by two other factors. One was the simple issue of funding. At the time of the *Apollo 11* success in the summer of 1969, there were enough additional Saturn V rockets, and plans in place, for nine more trips to the moon. The expectation was for lunar landings right up to *Apollo 20*. This sat very well with scientists who were anxious to transition from the mostly engineering emphasis of the first landing to those that would return reams of useful science. The safest lunar sites—the ones that were flat and therefore chosen for the initial landings—were also the least exciting, with the lowest potential for exhilarating discoveries.

NASA's Apollo plan dictated that the first few missions following the first, designated the H series, would use existing equipment that would let astronauts stay on the surface for up to thirty-six hours. This would be followed by the J-series missions, using significantly modified LMs that included a ground exploration vehicle or Rover. With these improvements astronauts would stay on the moon for three full days, and bring back more than double the weight of earlier lunar samples.

Unfortunately, Lyndon Johnson in his outgoing 1970 budget slashed NASA's moon funds so that there was no money whatsoever for anything more than a total of four landings. Richard Nixon, as incoming president, restored a paltry seventy-nine million, which at least let the program continue beyond *Apollo 14*, and, more important, kept the J series alive. But it was already starting to look like the final few missions might be cancelled.

The second big factor, not yet apparent when the three *Apollo 11* astronauts returned to cheering crowds, was the surprisingly deep public apathy toward later landings. Although NASA had already witnessed the media's fickleness during the Mercury and Gemini program, nobody was quite prepared for the paucity of press coverage given to the later Apollos. By the time of *Apollo 16*, landing on the moon no longer merited the front page or the lead story in the evening news. Only the tiniest percentage of the general public knew the names of the current astronauts who had vacated planet Earth, or had the slightest notion of that mission's objectives.

A third reason, this one discussed only out of media earshot, pounded the nails into the coffins of the final three Apollos: the public relations cost–benefit ratio. The cost in this case was not money, but the potential risk to the U.S. reputation. We had succeeded; we had sent people to the moon. Each visit incurred serious risk. If the program continued, sooner or later a set of astronauts would not return. Shouldn't the United States quit while it was ahead? If it stopped with, say, six or seven landings with no one hurt, the Apollo books would be closed on nothing but a glorious and positive note. But if one of the final flights resulted in disaster, the program would forever have an asterisk, a black mark next to it. American pride and reputation would suffer. Why take the chance? Why keep going until something went wrong? Stop soon, save the money, and move on to something else.

Well, Richard Nixon was not going to be the one to argue with any of these points. The three factors proved irresistible. *Apollos 18, 19,* and *20* were canceled. The astronauts who'd spent years training for these lunar visits were crushed, and researchers had to make each of the existing six landings count for all the science that could possibly be squeezed into them.

Meanwhile, in preparation for Apollo, the United States had launched a series of unmanned robotic probes to soft-land on the lunar surface during the mid-1960s. The most successful of these was Surveyor, and by early in 1969 it was decided to send the second Apollo to a touchdown a mere walking (or hopping) distance from the *Surveyor 3*. The main problem with this idea was that *Apollo 11* had landed nearly four miles from its planned target site. How, then, could NASA now pinpoint a landing to within a few hundred yards?

This very challenge was appealing to planners. It became the primary goal of *Apollo 12*. The hope and expectation was that navigation would be perfected by increased accuracy in monitoring the LM's position after it had separated from the CSM. This way, in the twenty minutes available before it fired its descent engine, last-minute tweaks could be made in its instructions, and it could begin its final descent more precisely.

It was a grand idea: the astronauts would stroll over to a U.S. robot lander on the moon, snip pieces of it for return to Earth, and analyze it to see what two-and-a-half years in the moon's hostile environment had done to the materials. It was a simple and dramatic goal that the public could easily grasp, that NASA's PR department thought would keep the moon missions on the front page.

The landing-walking astronauts for this odyssey were to be Pete Conrad and Alan Bean, with the orbiting CSM piloted by Dick Gordon. (Ever hear of him? Didn't think so.) November 14 was the launch day, and everything was in place except for some worrisome clouds. NASA's PR department was further excited by enhanced plans for live color TV coverage. They could not know that Murphy's Law was poised to operate—twice.

# The Strange Adventures of *Apollo 12*

It was a very different group of men that flew *Apollo 12* from the ones of Tranquillity Base. The difference wasn't one of competence—both groups were probably equally able. Rather, it was a matter of personality. The *Apollo 11* crew had been serious-minded characters, with Mike Collins alone given to slightly letting things "hang loose." Later, when musing about that first manned landing on the moon, Collins expressed surprise that such a historic shared experience hadn't bonded the three men. He said that there had been little personal warmth or genuine camaraderie aboard *Apollo 11*, nor did the trio hang out or become friends afterward. Not that there was any animus; rather, they had a job to do, and they did it—coolly, professionally. Collins was retrospectively disappointed that nobody let any of the others "in" to his private thoughts or feelings, and that none of the historic crew that first landed on the moon had in any way grown closer to any of the others. They were not friends. For his part, Aldrin said many years later that he had suggested an annual reunion, but it never happened.

The *Apollo 12* crew was completely different. This was a far more light-hearted bunch, who did feel deep and genuine bonding that accompanied the personal respect each felt for the others. On the moon they laughed and giggled and genuinely enjoyed the experience, and regarded each other as friends before and afterwards. Indeed, Conrad and Bean had known each other for many years, and were good buddies. Bean wouldn't even have been there if Commander Conrad hadn't made that specific request to Deke Slayton.

"We're good friends today," Bean recalled a dozen years later. "We just liked each other."

But their shared jolly adventure started out in a truly frightening way.

Just thirty-six seconds after the immense Saturn V lifted off in the morning of November 14, something went horrifically wrong. All the warning lights simultaneously lit up on the crew's console, and banks of breakers tripped. None of the men had ever seen any failure this comprehensive in their years of training, even in the many diabolical simulations that had been sprung on them. Down at Mission Control, entire banks of data turned blank.

Lightning had struck the ascending rocket.

The launch, allowed to proceed despite low-hanging clouds, had now met with near disaster that only the superb Saturn V engineering was preventing. Due to the roar of the engines and lack of windows, the men had no inkling that lightning had been the problem.

"We didn't know what happened because all we saw was a bunch of warning lights," Bean said afterward. "I didn't know how that many warnings could come on unless the Command Module had separated from the Service Module."

"Okay we just lost the platform, gang," he tensely radioed to the ground. "We had everything in the world drop out."

Hearts were beating rapidly on the ground as well as in the air. But backup systems were saving things, at least for the moment. Indeed, onboard batteries, never designed to run the whole show, kept the voltage up high enough to power the all-important navigation system so that the ship wouldn't careen out of control. But the guidance platform did briefly go out.

Nobody had ever seen anything like this, and everyone thought the order to fire the escape tower rockets, pulling the capsule away, would be given any moment. Since the problems seemed to be electrical, the flight director now looked toward his electrical expert, controller John Aaron, who was staring at his screen in amazement. He'd seen this pattern long, long ago, during a crazy simulation. Suddenly he remembered, and it came to him, what should be done.

"Tell them to switch the FCE to auxiliary."

"What?" asked the puzzled director. "What's that?" But he had the Cap-Com relay those exact instructions. The astronauts, still ascending in their power-starved Command Module, repeated the radioed suggestion and then precisely echoed the flight director's reaction:

"What? What is that?"

But Commander Alan Bean knew where it was—a never-used little switch, one of among a gazillion that occupied a panel over his shoulder. He flicked it to the AUX position, and all the power and gauges and lights and controls promptly came back on, just like that. (After this, John Aaron became a sort of instant legend in Houston, a stature that, oddly enough, would grow even greater a few short months later.)

The rocket continued upward with everything seemingly working, and the two remaining stages fired on schedule. Bean started giggling like a lunatic. Whether from sheer relief or the simple weirdness of it all, he could be heard laughing himself all the way into orbit.

Nobody observing it all could doubt that this was already the strangest spaceflight of all time.

All that remained was to spend much of the first orbit getting the guidance platform realigned, since its data had been completely erased in the incident.

There was a serious remaining issue, of which the astronauts as well as the controllers were painfully aware, but didn't dare discuss openly. The lightning had struck the nose of the ship where the parachute system was located. Inside were various explosive charges designed to eject the nose cone and fire the chute during the Command Module's descent to the sea. What if these had been fried? What if they would no longer operate? In that case the men would come in at over 350 miles per hour and be killed upon impact. There was no way for anyone to test or examine it.

Years later, musing about the situation, Bean explained NASA's reasoning in letting the men continue the mission with such uncertainty as to the ultimate outcome:

*Either the parachutes are going to come out [when Apollo 12 returns] or they're not. But we can't determine that. There's certainly no reason to rush back to Earth to see that. So we might as well send them out to the moon and let them do all that stuff, and when they make the regular reentry, if the parachutes come out, great. If they don't, well, they lived ten days longer.* \*

For now, though, once they were safely on their way, Pete Conrad told Mission Control, "I'll tell you, that's a terrible way to break Al Bean into spaceflight."

It was a jovial, spirited crew. At one point when Houston had a suggestion for more instrument tests, Gordon said, "Okay, sounds good. We really don't have any place to go tonight so we don't mind working late."

It was a smooth three days coasting toward the moon. Commander Conrad's style was to do things by agreement rather than unilateral command, and the men got along. The TV camera was an improved variety and viewers on Earth frequently got to watch the men work and cavort:

*Apollo 12*: ". . . all these things we didn't have in Gemini, like toothpaste and shaving. We are really having a ball up here."

Houston: "All dressed up and no place to go."

*Apollo 12*: "Oh, we're going some place. We can see it getting bigger and bigger all the time."

Then there were the strange phenomena. For example, the men could see an odd, distant tumbling object out the window that seemed to be keeping pace with them. They assumed it was the spent, ejected third-stage rocket. Pete Conrad radioed:

"We have had it ever since yesterday and it seems to be tagging along with us." He later added, "We'll assume it's friendly anyway, okay?"

Houston replied: "If it makes any noises it's probably just wind in the rigging."

\* From interview with Alan Bean in *Footprints* by Douglas MacKinnon and Joseph Baldanza (New York: Acropolis Books, 1989).

Houston did not tell the men that the strange "UFO" they kept sighting could not possibly be the S-IVB third stage the men assumed it to be, since the latter was nearly 3,000 miles away. It would have been as impossible to see from the CSM as a person in Boston trying to spot a school bus in Los Angeles. It was something else, but nobody ever found out what.

When the men were a half day from lunar orbit, Conrad delivered one of his only imperious orders. It shocked the other two. He ordered them to defecate, right then and there.

"I don't want anyone to have to take a crap on the moon," said Conrad. The others just stared at their commander, wondering if this was some sort of joke.

"We don't have time, we won't have time," he continued. "I want you guys to get down there in the lower equipment bay and take your clothes off and get your little plastic bags and I want you to crap now so you won't ever have to do it again until we're headed back home."

Bean later recalled: "We said, 'You're nuts, Conrad. We don't want to do it, we can't do it, we're not gonna do it.' So he said, 'Get down there!'

"We thought, 'Are you kidding us? I mean, you're gonna start regulating nature here?'"

But Bean says that they each tried, and succeeded, and it worked, because nobody did defecate again until they were on their way back home.

The *Apollo 12* landing site in the Ocean of Storms was more than 500 miles west of Tranquillity, sitting very near the moon's equator at just three degrees south latitude. When the two vehicles separated, and the LM's descent engine fired, and the final descent took them within two miles of the touchdown spot, Conrad couldn't believe his eyes:

"There it is. Son of a gun, there it is! Right down the middle of the road." Their guidance had been so accurate that they were heading to an automatic landing *too* close to the targeted *Surveyor 3* spacecraft, sitting in the middle of a wide flat crater. Conrad took over and piloted the craft beyond the crater to a seemingly smooth spot. The ground vanished in a cloud of dust, just as it had for Aldrin and Armstrong, when the LM was fifty feet high. At this point, he mainly used instruments to guide the craft vertically down, so gently that when he cut the engines when seeing the contact light, at a height of six feet, it touched down at a speed of only two miles per hour, about the same force that a person feels when stepping off the bottom tread of a staircase.

Gordon, orbiting in the CSM some sixty miles up, used a 28-power telescope to scan the surface for the LM and, amazingly, found it: "He's on the surveyor crater! About a fourth of the surveyor crater to the northwest!"

So far, things had gone exceedingly well. The men had landed just 500 feet from the *Surveyor 3*. They had proven that pinpoint lunar landings were possible—and this demonstrated accuracy was what allowed future Apollos to venture into hazardously rough terrain.

Conrad backed out of the LM, went down the ladder, and was surprised to see how far the final rung was perched from the ground. He was one of the shortest astronauts, and his legs simply wouldn't reach, so he simply leaped off. Then he told Houston: "Whoopie! Man, that may have been one small step for Neil, but that's a long one for me!"

Unfortunately, their good luck was about to take a turn. One of the crew's first tasks was an important procedure to benefit the viewers back home: set up the improved color TV camera with a high-gain antenna that would let the world enjoy live lunar images with amazing clarity. Conrad set up the umbrella-like antenna, went to position the camera on a tripod, and then a moment's thoughtlessness: he accidentally pointed the camera at the blinding sun. It was just momentary, but that was enough to fry the vidicon tube. It was dead, beyond resurrection. There would be no images from the moon.

The experts at NASA's PR department slapped their heads and groaned. Without images, the astronauts' time on the moon would be reduced to staccato audio comments, and the impact for those at home would be greatly diminished. Thus it was that the trip to the *Surveyor 3* and all the rest wound up being a media dud, and TVs around the world switched to movie channels. It truly was disastrous for NASA's efforts to keep media attention at a high level.

*Apollo 12*'s science return was another story. The men set up the EASAP science package with its seismometer and other equipment, and busied themselves with sample collection and a solar wind experiment. They wound up gathering seventy-five pounds of moon rocks, which proved to be wholly different from the samples from the Sea of Tranquillity.

The transmitter on the *Apollo 11* science package had failed after just a few weeks on the surface, and the improved *Apollo 12* version was expected work far better. (*Much* better, it turned out. Designed to live for a year, it was still working and transmitting flawlessly eight years later, when funding for its continued operation was cut in 1977 and the equipment merely switched off. But by then its seismometer had successfully registered 2,300 tiny moonquakes and various-sized meteor impacts, as well as the large vibrations produced when each of the unneeded LM modules were sent crashing onto the surface.)

The men returned to the LM and rested. An eight-and-a-half-hour sleep period was scheduled, but they were too keyed up to rest for more than a couple of hours, and finally, like insomniacs looking for something to do in the middle of the night, asked and received permission from Houston to prepare for their second EVA.

This was the long one—involving a full mile of walking, and venturing a distance so far from their LM (1,300 feet) that it would sometimes vanish behind low rises, due to the dramatically closer horizon of the lunar surface. On Earth, the horizon appears 2.7 miles away to the eyes of a person standing five-and-a-half feet tall. But on the moon, the horizon is less than a mile away: low objects can easily vanish from view before one has wandered very far at all.

Conrad and Bean took turns losing their balance and falling, a strange sensation because hitting the ground is just as inevitable as it is on Earth, but the process takes six times longer, so one has much more time to be aware of the helpless tumbling sensation. In each case, the other astronaut would come over and help the fallen comrade up. Although an astronaut can lift himself from the surface unaided, the bulky spacepack makes it easier if he gets a hand. Both men reported that it was fun to see how a mere finger can effortlessly help lift a full-grown man in the moon's weak gravity.

Conrad and Bean walked around the crater to *Surveyor 3*, retrieved its TV camera, cut samples of its cables, and snapped pictures of the way its footpads had obviously hopped a bit in the powdery surface when it landed. The biggest surprise of the mission came much later when this material was carefully analyzed.

A germ, a bacterium, was found in the foam in the TV camera. And the alpha hemolytic *Streptococcus mitis* was alive! It had survived on the moon's airless surface for thirty-one months, in the vacuum of space, enduring cold of minus 260 degrees Fahrenheit and heat of over 150 degrees. Obviously the foam had been contaminated prior to its journey into space even though it was supposed to be sterile. The germ's unbelievable longevity demonstrated, if any proof was needed, why it's so hard to shake a strep throat.

*Apollo 12*'s return was *almost* uneventful, which was very good news considering the serious worries about the possibly lightning-damaged parachute system. Yet, just when the trip was seconds from being officially over, one of the men received the worst injury that would ever be suffered during any mission to the moon.

At splashdown, the force of the water impact dislodged the 16-mm movie camera that had been attached to the CM wall with a bracket. The massive camera came down on Bean's head, knocking him briefly unconscious and opening up a wound that needed six stitches to close.

"It cold-cocked him," explained Conrad afterward. "He was out to lunch for about five seconds, staring blankly at the instrument panel. I was convinced he was dead over there in the right seat."

With helicopter pickup and TV cameras awaiting their imminent appearance on the aircraft carrier USS *Hornet*, the crew stopped the bleeding and hastily hid the injury under a large Band-Aid.

*Apollo 12* was a great success that further raised the optimism level at NASA. Perhaps that's one reason why, five months later, nobody was quite prepared for the shock of *Apollo 13*.

○

# CHAPTER 20

# Triskaidekaphobia:
# When Bad Things Happen
# to Good Spacecraft

*The moon is nothing*
*but a circumambulatory aphrodisiac*
*Divinely subsidized to provoke the world*
*into a rising birth rate.*
                    —Christopher Fry, "A Sleep of Prisoner"

A *pollo 13*'s unlucky flight is now very familiar, thanks to Jeffrey Kluger's excellent book and the gripping Tom Hanks movie. Since it is not our purpose to retell stories already widely circulated, we'll instead focus on this ill-fated mission's quirky, little-known aspects.

The story naturally begins with the name. Just as many office and residential buildings omit the thirteenth floor, there were indeed discussions about whether there should even be an Apollo named *13*. Such debates were brief: of course there would be. NASA was the world's most prestigious, widely known scientific agency. If *they* wouldn't put a firm stop to superstitious nonsense, who would?

Good call, bad result. The explosion that crippled *Apollo 13* did little to cure the world of triskaidekaphobia. Nor was it helpful that it happened on April 13.

The mission started quietly enough; in fact, too quietly. Gone were the packed crowds that had come to witness the launch of *Apollo 11* just nine months earlier. Less than 10 percent of that number now attended, even though this April 11 launch was at a more pleasant season and a more convenient time of day. Vanished were the copious preflight press interviews. Even Houston TV stations refused to interrupt their regular daytime programming for live coverage of the liftoff, at exactly thirteen minutes past 2:00 p.m.

Commanded by Jim Lovell, who had logged more time in space (572 hours, mostly aboard Gemini) than any other human being, the Saturn's awesome first stage performed flawlessly as always. But when it came to the nervous, pogo-prone second stage, its center J-2 engine suffered a pressure drop that caused detectors to entirely shut it down prematurely, a full two minutes, thirteen seconds early.

"Houston, what's the story on engine 5?" radioed Lovell, his crewmates, Fred Haise and Jack Swigert, looking on with concern.

Those two wouldn't even have been in their seats here on *Apollo 13*, but for a bizarre set of circumstances. Fred Haise had been the astronaut bumped from the first moon landing flight when Mike Collins was inserted as CM pilot. Finally, Haise's chance to go to the moon had come.

As for Swigert, he was a last-second replacement for Thomas Mattingly when, just five days earlier, the entire crew was exposed to measles and tests showed Mattingly the only crew member who was susceptible to the disease. There had never been a substitute this late in the game. The whole thing was screwy.

Now the motley crew watched as the outer four J-2 engines tried to compensate for their companion's silence by remaining on for an extra forty-four seconds. It wasn't quite good enough. When the third (S-IVB) stage ignited, it had to remain lit for an extra nine seconds, and now they were close to the original orbit, though they had arrived nearly a minute late.

It wasn't a great start, and they had less fuel remaining than any other Apollo at this point, but they were eventually cleared for trans-lunar injection and their ongoing trip to the moon.

What neither they nor anyone on Earth knew, was that they were carrying a time bomb aboard.

Large tanks of liquid oxygen lurked below them in the Service Module, and each required periodic stirring and temperature adjustments, accomplished by vanes, fans, and thermostats. Years earlier, NASA had changed over this electrical equipment from a 28-volt DC system to a superior 65-volt version, and the contractor, Beech Aviation, had duly installed the new higher-voltage items. Unfortunately, Beech had neglected to tell their own supplier to substitute *switches* suitable for the higher voltage. And so it was that switches that could overheat or even lock up, that could cause the surrounding insulation to burn, existed in the high-pressure oxygen tanks— the very environment that would most promulgate fire or explosion.

A deadly explosion could have occurred in any of the previous Apollo missions, but it took a final straw to ignite it. One of the oxygen tanks for this flight had been dropped by a crane a year earlier. It had only fallen two inches, and seemed to be all right. But the extra testing that it then underwent, and the draining of oxygen inside, necessitated the faulty fans and switches being used more often than on any previous flight, and the wires inside the tank had already started to fracture.

No one, of course, knew any of this. But at two days and six hours into the flight, only a day from reaching the moon, just as the men signed off from one of their live TV shows that very few on Earth were bothering to watch, NASA asked them to perform one small final chore before hunkering down for their sleep period.

"*Thirteen*, we've got one more item for you when you get a chance. We'd like you to stir up your cryo tanks."

Jack Swigert was on duty near the switches. "Okay," was all he said.

Telemetry received by ground controllers showed power going to the oxygen tank but suddenly, there was a whopping 11-amp spike, a short circuit. Wires now started to burn inside the oxygen tank, and the tank's pressure began to rise. It was a bomb, nothing less. Then came a huge 22-amp surge, followed by the tank pressure reaching the upper limit of the monitoring gauges, along with sharply escalating temperatures. This data was only examined later; at this moment the only thing the crew was aware of was a sudden, tremendous explosion, and the violent shaking of the entire craft. There was also a momentary loss of radio communications as the exploding wall of the service module below them knocked the antenna from its alignment with Earth.

The crew had no idea what had caused the explosion. If they had, they would have been appalled at the immense damage to systems vital to their survival: an entire side of the cylindrical service module had been blown away beneath them. But they would have been simultaneously relieved that the expanding fire below them had now been promptly snuffed out by its sudden confrontation with the vacuum of space. Fred Haise, in the tunnel leading to the LM, now backed out. Jim Lovell, in the lower equipment bay, raced up. Jack Swigert pulled himself into the right seat, where he could monitor the ship's systems. He called Mission Control: "Okay Houston, we've had a problem here."

Houston: "This is Houston. Say again please."

Swigert: "Houston, we've had a problem. We've had a main B bus undervolt."

The problem of course was much worse than that, but it was a drama whose full consequence was to unfold slowly, in the course of the next hour.

Looking at the gauges, it was clear that weird things were still happening. First, the ship's thrusters were firing almost continuously. They were trying—and succeeding—in keeping the spacecraft from tumbling, because some strange torque was being continually applied into space. When the men looked out the windows, they could see the cause: material spewing out of their ship below them in the Service Module, spraying energetically into space. It was the oxygen. Not only had they lost everything from tank two, but the explosion had irreparably damaged tank one, which was now slowly emptying its contents as well.

The expanding bubble of oxygen mist was picked up by a Canadian telescope—the farthest distance a man-made object had ever been detected from Earth.

With one tank gone, the landing would have to be canceled, and the men were immediately crestfallen at the certain knowledge of that fact. But with two tanks gone, their very lives were in peril. In fact, there was no real way they should survive at all. The oxygen was not merely used for breathing, it was used, when combined with the tanks of hydrogen, for creating electricity and water. With no electricity there was no returning. Their batteries would not last long enough.

An hour and a half after the explosion, the pressure in tank one had dropped from 1,750 pounds per square inch to just 200 pounds, and it was clearly heading down to zero. This is when Houston first said:

"We're starting to think about the LM lifeboat."

And *Apollo 13* answered: "Yes, that's something we're thinking about, too."

The LM lifeboat. This was *not* a sudden, unprecedented notion, contrary to the *Apollo 13* movie. It had been periodically discussed in emergency planning. After all, the LM had its own independent oxygen, electricity, and the staples to keep two men alive for up to thirty-six hours. The problem, of course, was that now there were *three* men, and the soonest they could get home would be another 100 hours. Was there any way to stretch supplies for two men and 36 hours into three men and 100?

They had to try. Already, the only power in the Command Module was coming from the steadily draining batteries. In a few short hours, everything would go dark anyway.

Meanwhile the next shift had reported to Mission Control, including electrical controller John Aaron. He quickly sized up the situation, and then expressed amazement that *Apollo 12* was still being allowed to run on its batteries.

"Get them into the LM, and turn off all the power, all of it, everything," he ordered. He knew that every volt of current would be needed later in order to bring them back. Why was it being wasted now?

If they could immediately shut down absolutely everything in the Command Module, so that the batteries would still retain some life, they could use it four days hence to briefly reignite the module's systems for a reentry through the atmosphere. In the meantime, they could crawl into the LM and simply try to survive. Such a cold shutdown and, later, an attempted power-up from a totally cold state had never been tried before, never even practiced before. But what choice did they have?

The SPS engine was presumably dead and useless, but the LM still had its unused descent engine. Controllers at Houston and engineers and specialists throughout the country, at the many subcontractors who'd built the LM and CM and Service Module and its myriad subsystems, came in and frantically worked out the math for an engine firing needed to make the craft loop around the moon, pick up speed, and be hurled back to Earth—and to stretch the existing resources to allow the men survive for four days. As it was, if nothing was done, they would indeed pass behind the moon and be hurled Earthward, but they'd miss their home planet by 50,000 miles. No cigar.

There were two possible life-saving scenarios. One: Eject the service module now to lighten their mass, which would allow an LM engine burn around the moon that would bring them back a day and a half early. But this would subject their CM's heat shield to days of icy temperatures in the cold of space, and nobody knew how the all-important materials would fare under those conditions.

Choice two was more conservative: Leave the Service Module attached until the usual time for ejection in the vicinity of Earth, fire the LM engine, and return one day early. This would preserve the shield, but the oxygen and electricity would be very, very tight.

By now the denouement of this nail-biter is well known. NASA's contractors and subcontractors worked out how to fit the "square hole" CM lithium hydroxide carbon-dioxide removal canisters to the "round hole" of the LM's equipment using duct tape and cardboard, and solved many other problems as well to help the men survive.

Survive yes, but not comfortably. Lights, heaters, everything not absolutely vital had to be shut down for the four days, with electrical consumption turned to just 20 percent of normal. Temperature in the LM plummeted to barely above freezing. The men shivered uncontrollably, unable to sleep.

While the world anxiously watched and listened (they were paying attention *now*), the ship whipped around the moon's far side a mere 158 miles above the surface. Poor Fred Haise, bumped from *Apollo 11* and now bumped by bad luck from getting down on the surface, could only stare through the window and take pictures. With the eventual cancellation of *Apollo 18, 19,* and *20,* neither he nor Jack Swigert would ever again get this close to the moon.

The Houston and contractor problem-solving was continuous. Even the act of slowly rotating the joined spacecraft to keep from getting too hot on one side, too cold on the other, presented challenges. In a normal flight, the thrusters are on the CM, near the center of mass of the two linked craft. Now, using only LM thrusters, all the torque was being applied from one end, and it tended to twist the joined craft to near the breaking point.

Inside, the men were increasingly miserable. Beyond the psychological torment of not knowing if the explosion had damaged their unseen heat shield, their physical comfort worsened by the hour. Water was rationed to six ounces a day, and the men quickly became parched and dehydrated. The food packs, designed to be reconstituted with hot water, would not work with cold, and cold was all they had. The systems could not remove all their exhaled moisture from the air, and as the interior became ever-colder it also became increasingly humid. The windows fogged up, and water started condensing on all surfaces. More alarmingly, the countless electrical switches had become dripping wet, too.

Fred Haise succumbed to a raging kidney infection; his fever shot to 104 and he shivered uncontrollably, utterly unable to sleep. The men attempted to use the CM's larger space for sleeping, but with lights out and thirty-something-degree air, "it was like a tomb."

There was no way to vent human waste overboard because Newton's reaction/opposite-reaction law meant that the ship would respond with motion that would have to be corrected by thruster firings that they could not afford. The men later recalled: "We had urine all over the place. What to do with it taxed our ingenuity." They found imaginative storage places, and used moon rock bags.

More depressingly, the men had serious doubts that they would make it back alive at all. Finding the all-important guide stars for navigation was impossible because the ship was surrounded by thousands of glittering floating particles, the debris from the explosion that was traveling along with them. Fred Haise radioed: "Looks like I'm in the middle of the Milky Way." Controllers worked out ways to use the sun instead.

In addition to stretching out the life-support resources, there was the issue of power management once the Command Module was turned back on. The batteries only had about two hours of life, and even then not everything could be turned on. All the pre-entry work would have to be choreographed like an orchestral piece, and it would have to be done quickly and accurately. John Aaron met with controllers in charge of numerous systems, and they each insisted that the CM absolutely had to have one or another electrical system working for the reentry. Those in charge of recovery urged him to let them turn on the locator beacon, so that ships could find the capsule once it hit the sea.

Aaron was ruthless: "If I can get them back to even reach some ocean, you can find some way to locate them in the sea without the beacon," he explained. The beacon stayed off.

For all that, the CM batteries were still alive when they were needed four days after the accident, the wet electrical switches did *not* short out, and the heat shield was not critically damaged despite the oh-so-nearby explosion in the Service Module. It could have been much worse. If the blast had occurred a single day later, all the men would have died because the LM "lifeboat" would then be on the moon, its descent engine and provisions spent. If the problem had afflicted *Apollo 8*, which carried no LM, or happened to *11*, *12*, or *13* on their way *back* from the moon, the men would have been goners—and a slow death at that. There simply would have been no unused LM to use as a lifeboat, and all the oxygen, $CO_2$ scrubbing, and electrical facilities would have been long expended before reaching the safety of Earth.

Ironically, one thing did go spectacularly right. *Apollo 13* came down closer to their planned aircraft carrier than any other mission, although the men were so weak and severely dehydrated that they could barely walk on the deck. Lovell had lost fourteen pounds and Haise required three weeks in a hospital.

But they were alive.

○

# CHAPTER 21

# Postmortem

> *. . . in its box of sky*
> *lavender and cornerless,*
> *the moon rattles like a fragment of angry*
> *candy.*
>
> —e. e. cummings, "Tulips and Chimneys"

orbid interest? Awakened awareness? Whatever the reason, the launch of *Apollo 14* nine months after the hair-raising *Apollo 13* accident rekindled the public interest in the moon program. No longer was a safe flight guaranteed in the public mind. Now there was danger, and danger meant excitement, and excitement brought out the crowds like those at a NASCAR event. They lined the Cape Kennedy causeway on the last day of January 1971.

*Apollo 14* was set to venture to the lunar region meant for *Apollo 13*, the Fra Mauro highlands. The year 1970 had come and gone without a single person stepping on the moon, so this would be one of those "we're back in business" flights, the way *Apollo 7* had been. It would be piloted by none other than Alan Shepard, the very first U.S. astronaut into space; he'd flown the initial manned Mercury in its suborbital flight of May 5, 1961.

As a member of the Original Seven, and the astronaut who had garnered the very first heady idolization, the ticker-tape parades that erratically greeted—or were withheld from—various astronauts during the 1960s, he was itching to return to space.

It wasn't easy. Shepard had been grounded when he developed Meniere's disease, a serious inner-ear affliction that caused unpredictable vertigo. Unlike the wing clippings for extremely iffy or minor ailments, as in the case of Deke Slayton, Meniere's disease was sufficient to ground even a cub student pilot. But Shepard had undergone a new kind of delicate operation that had fully cured him, and he won back his wings. His status as national hero and *Mercury Seven* astronaut easily got him a place on a moon mission even though, at forty-seven, he was far and away the oldest astronaut.

He was a bit of a loose cannon, too, in that his superb piloting skills were somewhat offset by his "don't bore me with this crap," seen-it-all veteran's attitude toward the voluminous intellectual and science instruction he was forced to undergo as part of his Apollo training. He was already a wealthy man, having displayed high business acumen, and now, pushing fifty, he really didn't need to do this at all. But pass up a trip to the moon? No way. Yet his grizzled-veteran-who-knows-the-ropes attitude would come back to haunt him, and when it was all over, *Apollo 14* would return the least useful science of any Apollo mission. In hindsight, many in NASA later wished they'd picked someone else to command number 14.

But the priority after the failed *Apollo 13* was success, and if you mostly needed an amazing test pilot to have the best chance of getting to the moon and back, Shepard was the right guy. He was right up there with Chuck Yeager, having flown dangerous experimental craft at Mach 2 to 90,000 feet way back in the mid-fifties. As for his age, there was indeed open questioning about choosing a man approaching fifty, and even on the day of the flight, when getting ready to enter the CM, his crewmates mischievously handed him a walking stick. He shot them an unamused look.

Those jokesters that would accompany Shepard were Edgar Mitchell, who would join him on the surface as pilot of the LM, named *Antares* for this mission, and Stuart Roosa, who would command the CM *Kitty Hawk* and remain in moon orbit. Between the three of them they had had a grand total of fifteen minutes above Earth's atmosphere—all of it from Shepard's suborbital arc a full decade earlier. Some called this "the rookie flight." Never before, and never since, did Apollo fly with such a space-inexperienced crew.

The Saturn V fired on the afternoon of January 31 into a murky drizzle, seemingly violating the "never again" rule after *Apollo 12* had been hit by lightning. But this time, special planes patrolling the area had detected no electrical activity in those clouds and, after a countdown was halted, it proceeded again with all fingers crossed. The controllers watched the deafening six-story rocket immediately vanish into low clouds. But the ascent into orbit was uneventful. The only problem came later, when the crew had serious difficulty snagging the LM from its garage. The latches just wouldn't grab despite repeated attempts at ever-faster speeds. The mission was now in jeopardy. Finally Houston gave the go-ahead to try for a "hard dock" instead of the usual initial three-latch soft dock, and this time it worked. And the rest of the three-day coast to the moon was uneventful.

The landing, however, was not. At first the landing radar on the LM went on the blink, but the men continued downward anyway, assuming that they'd be able to do it manually if they had to. This would have been against mission rules, and there was widespread later speculation—confirmed by Mitchell—that the pair intended to keep going even if they got an order to abort and to fire their ascent engine.

"We would have pulled our earplugs," said Mitchell years later, meaning they would have pretended not to hear the command to terminate the landing approach. When Shepard was later asked if he would have disobeyed orders and kept descending, he would always give the same reply: "Next question."

But it never came to such a mutiny, because the landing radar came back on and the approach proceeded normally.

This would be the final H-series mission before the newly renovated LM came into service, the last one before use of the spanking new lunar rover, yet scientists hoped for serious science returns. In addition to the usual ALSEP package of experiments, there was now an atmospheric gas experiment, and an active seismic experiment.

The latter consisted of a long cable that Mitchell laid out on the lunar surface. At exact 150-foot intervals he set down vibration detectors, then fired what he called a thumper at specified intervals. This was a flat metal plate on a pole that created hard poundings of the surface, that would be measured by the various detectors to give clues about the subsurface structure. The astronauts also set up a grenade launcher that later (after the men were safely back to Earth) fired four explosives for even deeper subsurface analyses.

But the headline event was the astronauts' second EVA, which was supposed to take them to the rim of an interesting deep feature: Cone Crater. It was a long distance away, and while going there they were supposed to be "assisted" by a rolling instrument cart called MET that would help them transport the tools they needed on the mile-long walk. In reality, the rickshaw-like cart proved difficult to roll in the deep, powdery, undulating surface, and the men wound up simply carrying it between them.

That second EVA was the low point of the mission. The men struggled uphill, searching for the rim of the crater, and here is where Shepard's impatience with "intellectual details" (others would say, a lack of thorough preparation) came back to haunt him. After two hours, their hearts racing from the effort, the astronauts had to acknowledge that they were lost.

They did not know where they were in relation to the crater. Meanwhile, Shepard, periodically cursing, had trudged uphill collecting samples while scientists watching from Mission Control muttered disapproval at how little specific data he was bothering to annotate about each one—which would limit the geologic usefulness. Finally the men were ordered back. Later analysis showed that they were a mere fifty feet or twenty-five steps from the crater rim, but had not recognized it. On the moon, it is extremely hard to judge distances. Unlike Earth, where far-away vistas appear distant because the intervening air imparts a light blue cast, the lack of lunar atmosphere means that near and far look equally sharp and stark. Moreover, shadows are inky black, and the overall effect is one of heightened ultrahigh contrast between sunlit portions and shadowed portions of any scene. Still, NASA's own geologists later thought it likely that greater preparation on the part of the astronauts might have given the Cone Crater excursion a better outcome.

After this failed EVA, before stepping back into the LM, Shepard removed a golf ball he had secreted in his space suit, took out the handle of a geology instrument, and announced that he would now play golf. His bulky suit prevented him from using both hands, and his first one-handed attempts missed. But the third try connected, and he announced: "There it goes, miles and miles and miles." He hit a second ball, too, but later, more sober analysis showed that the two trajectories were nowhere near the "miles and miles" category. The first probably sailed 200 yards, the second twice that far. Still respectable.

Edgar Mitchell did an even stranger thing—much stranger. By prearrangement with four people back on Earth, he used his own leisure "off time" during the mission to perform ESP experiments using a series of twenty-five cards with symbols. The idea was to see if mental telepathy could work between Earth and deep space. The result? This depends on whom you ask.

Since 40 correct hits (out of 200 tries) would be expected by chance, and three of the four subjects scored lower than this, with only the final subject scoring 51 out of 200 correct, no objective scientists would declare this anything but a random result, a failure to demonstrate ESP. Moreover, the launch's forty-minute delay meant that all four subjects, who had been coached to expect ESP transmissions at specific times in the mission, but who were now operating on a wrong schedule, were now attempting to receive Mitchell's thoughts at a totally incorrect time, when he wasn't even "sending" them.

None of that seemed to faze Mitchell's later analysis, whose post-Apollo life was very much involved with psychic research. He characterized the results as "statistically highly significant." At any rate, Mitchell had kept the entire project secret from both NASA and his crewmates. He later explained that NASA's feelings about such psychic research was "totally negative. Totally closed"—an appraisal that surprised no one.

The astronauts blasted off from the moon, using an improved ascent technique that let them meet the orbiting CSM on the very first orbit. All in all, and except for the disappointment of the Cone Crater EVA and Shepard's subpar geological work, *Apollo 14* was a success.

Unfortunately, public and media interest in the odyssey started to seriously wane as the mission's days wore on, and few watched it on TV. In New York, a massive power blackout occupied its citizens, and elsewhere around the world media attention was focused on other events. No parades were held for the returning men and no congratulatory invitations to visit the White House were issued. Indeed, the *Apollo 14* astronauts didn't even receive a phone call from president Nixon, who was, as we've seen, more a lover of publicity than of Apollo. Nixon did phone Deke Slayton in Houston when the mission had ended successfully, and asked him to "pass along" congratulations to everyone involved.

The handwriting was on the wall. The moon missions had now become routine. From *Apollo 14* on, the men would continue to do live TV broadcasts from space and from the moon. The only problem: nobody was watching. Indeed, they couldn't watch even if they wanted to. The networks simply weren't carrying it.

NASA now made it official—there would be just three more trips to the moon instead of six. There wasn't money to keep going, and the public was apathetic, and (though they never stated it openly) NASA would quit while it was ahead, before someone got killed.

All understandable. Still, though few paid attention, the final three moon trips—the new J-series missions—were far and away the most productive, and they would land in the most dangerous and rugged parts of the lunar surface. Finally, engineering was going to start taking a back seat to science and adventure.

○

# CHAPTER 22

# "Can't You Americans Go Anywhere without a Car?"

*We are not very shy;*
*We're very wide awake,*
*The moon and I.*

—Sir William Gilbert, *The Mikado*

It didn't quite have the glitz and glamour of an auto show. But a new age of exploration began in the middle of 1971, when the United States took its obsession with cars into outer space.

Although the media and public were largely oblivious, the all-terrain vehicle brought to the moon was merely part of NASA's long-standing plan. Or, to explain it in NASA's own beloved technospeak, Apollo was now officially switching from its initial H series to the far-more-advanced J series. (There was also supposed to be an I series, consisting of flying lunar explorers, but this had long ago been scrapped.)

If the H series (*Apollo 11, 12,* and *14*) successfully proved that moon landings could be accomplished with precision, the J series would now "get real," with a new LM version that boasted twice the capacity of the original. It would double the previous version's on-surface carrying weight for supplies, and allow twice as much time on the moon itself (three days versus the original one and a half).

The new missions also featured a new space suit that would provide far greater physical flexibility (the men could now bend at the waist, for example). This allowed the astronauts to actually assume a sitting position, a necessity in order to drive the Lunar Rover Vehicle, or LRV, the most widely reported aspect of the "new" moon program.

Powered by two 120-amp batteries and 36-volt electric motors on each wheel, with batteries that could carry the rover to the impressive drive-out-of-the-city distance of fifty miles, the LRV could negotiate rough terrain and allow the crew to explore vastly increased areas at a time. It weighed just a quarter-ton, or 455 pounds to be exact, but could carry twice that weight.

To give the astronauts the vital life support they'd need during these extended EVAs, their new life-support system let them stay outdoors for up to seven hours at a stretch. Their packs now supplied them with drinking water and artificial orange juice, and even an internal Velcro patch so they could, if necessary, scratch their noses.

But all of this (except for that Velcro patch) meant significantly more weight, and here is where the great Saturn V's middle engine proved its worth. Without the full 7.5 million pounds of thrust, the extra weight of the new rover-laden LM, including a larger descent engine to slow this extra mass softly to the surface, would never have been possible.

The first crew to try out this brand-new equipment was the *Apollo 15* commander, Dave Scott, along with Jim Irwin to pilot the LM (named *Falcon* for this mission), and Al Worden as CSM pilot. Another very exciting facet of *Apollo 15* was the chosen landing site. It would be the most northerly of all the Apollo sites, in a narrow valley between mountains, bordered by the deep seventy-mile-long chasm known as the Hadley Rille. The latter, a dramatic Grand Canyon–like feature that averages as deep as the Empire State Building is high, has walls up to a mile apart. This site was the first major departure from the flat play-it-safe locations of the first three (especially the first two) Apollo landings.

*Apollo 15* blasted off on July 26, 1971, the heaviest Saturn V ever. While the flight to orbit was normal, an electrical problem kept the crew busy troubleshooting on the way to the moon. Also en route came an unusual procedure: previous Apollo astronauts had reported seeing flashes of light repeatedly zipping across their ocular fields of view. Eyes open or shut, it made no difference. Early crews, following well-known pilot wisdom, said nothing lest doctors or, more worrisome, psychiatrists would get involved. But when confidential conversations among the astronauts showed that *everyone* had seen the same lights, they were clearly neither hallucinations nor individual medical quirks.

But what were they, then?

On this flight all three crewmen were asked to wear opaque hoods and face the same way for nearly an hour, and report whenever they saw such a "meteor" flash across their visual field. All in all the crew reported fifty-four such sightings. Although the answer is still not definitive, the widely accepted theory for this strange but universal in-space phenomenon is that cosmic rays whipping through the astronaut's skull touch the optic nerve and cause a spurious signal to be sent to the brain. Cosmic rays were far more prevalent for Apollo astronauts en route to the moon; these things were not sighted by astronauts in near-Earth orbit (or later on the Shuttle or Space Station), where the men were within Earth's protective magnetosphere or magnetic field.

The trajectory to the moon was a new one, too, because *Apollo 15* needed to orbit the moon at a unique angle to the lunar equator. Indeed, once in lunar orbit, they would look out the window at features not seen by any previous visitors.

The descent burn was also longer than previous ones, due to the need to slow down a heavier mass, and the men were thrilled, flying feet first and eyes upward, as the computer aboard the LM automatically carried out the first part of the descent, skimming their craft right over mountain tops before entering the valley. This was the heaviest LM by far, with lower fuel margins, and the pilots were trying to resist the tendency of previous landings to stop and hover at about 150 feet up before backing slowly vertically down. Instead, when they took over from the computer at 700 feet (seventy stories above the ground) they attempted to fly a smooth continuing profile right down to where the ground contact lights would come on, four to six feet up. They succeeded, but hit with such a tremendous impact that Irwin later said "I was sure something was broken. We both just froze as we waited for [Houston controllers] to look at all of our systems."

But everything checked out okay. They had actually set down beautifully, a mile from the edge of the Hadley Rille, within glorious sight of the 14,800-foot-high Mount Hadley. It was a magnificent place, and the men— fully trained and excited by geology and fired up to get going—were to use their three days to full advantage. This was a sharp contrast to the seeming science-lackadaisical attitude of the previous Apollo commander.

The men had no trouble lowering the LRV, using dual ropes, and it was ready to roll fifteen minutes later. On that first drive the men "opened 'er up" to the LRV's top-level speed of seven miles per hour, about the same as a person jogging, and drove right to the dramatic edge of the canyonlike rille.

The LRV had an onboard computer that could automatically retrace its path if needed, and find its way back to the LM. This was an important safeguard: getting lost on the moon would be considered a serious error. Although the rover was capable of going more than fifty miles, mission rules limited the maximum straight-line distance from the LM to six miles, regarded as a distance that the men could reach on foot if the vehicle broke down.

Altogether Scott and Irwin drove six-and-a-quarter miles on that initial EVA, before returning for the first night's sleep. The next day, August 1, the astronauts took the LRV to the base of Mount Hadley. It was here that they found a white crystalline rock that their lengthy geological studies immediately told them was older than anything else around. The press quickly dubbed it the "Genesis rock," and later analysis did indeed date it to the birth of the moon itself, some four billion years ago.

This drive sloped uphill, until the men were on a rise 500 feet above the plain on which they had landed. Indeed, there in the distance three miles away, they could just make out the LM, in the valley between the gently sloping mountains. This day, the men covered a round-trip distance of eight miles. In between, they set out the science experiments near the LM, and kept meticulous notes about each rock sample they acquired.

Back on Earth, interest in the LRV created a temporary spike in press coverage. One Brit, reading a newspaper in India with a U.S. citizen peering over his shoulder, turned and asked: "Can't you Americans go *anywhere* without a car?"

On the final day's EVA, Scott and Irwin drove along the edge of the rille. But time constraints prevented them from exploring further, since it had taken far longer than anyone had imagined to pound in and remove a ten-foot-deep core sample. Although the moon's top few inches are powdery, the rough-edged material increasingly packs with depth at all sites. Even six inches down, it has a nearly rock-hard consistency that defies all easy efforts at penetration.

The men also left behind a plaque honoring the deceased astronauts of both the United States and Russia, along with a small metal sculpture that they lay prone in the lunar soil, to commemorate "the fallen astronaut." Then, before the men stepped back into *Falcon* for the last time, they carefully parked the rover and aimed its TV camera so that it would have a good view of the liftoff. The camera was remotely controlled from Houston, and would pan upward as their LM ascended. All told, the men had driven seventeen miles along the surface, which doubled the distance of all previous missions' explorations combined.

The next day, now safely orbiting the moon before firing their SPS to return to an Earthbound trajectory, the CSM released a satellite to orbit the moon; it returned information for more than a year, after becoming the first satellite ever launched from a manned spacecraft. The men also released *Falcon* to crash to the lunar surface, where seismometers from all the Apollo landing sites recorded the results from this impact of a known force.

The following day, now en route home, CSM pilot Al Worden performed a thirty-eight-minute spacewalk to retrieve the exposed 16-mm film from his craft's outside camera. Doing so, he became the farthest man from Earth ever to walk in space. Also that day, the crew performed yet another test of the visual flashes they were all experiencing.

It was a splendid mission, though the splashdown on August 7, 1971, was not quite routine. One of the three main parachutes opened fully but then immediately collapsed, its lines having been severed by some errant propellant. Observers on the carrier USS *Okinawa* radioed the module: "*Apollo 15*, stand by for a hard impact."

But it wasn't much more than a sharp jolt, and the mission was safely over. They had been off the planet for twelve days, seven hours, the longest to date.

Unfortunately, *Apollo 15* was to have a strange footnote. Unbeknownst to everyone when the mission ended, but revealed publicly a year later, all three men had taken along 400 special stamps that were passed afterward to a dealer in Europe. (The crew each received $8,000 from the dealer, who sold them for a $150,000 profit—about a half-million dollars in today's currency.) The three astronauts had arranged it so that the money would go into a trust fund set up for their children, which violated NASA's standards of conduct. Even worse, the men had also profited from a deal to manufacture and sell replicas of the "fallen astronaut" figure they had placed on the moon.

Congress angrily investigated, NASA was tremendously irritated, and all three men had their flight status revoked, even though they returned the money. Scott was effectively demoted to "special assistant on Apollo," while Irwin simply retired to found a religious organization. Worden moved to Ames Research Center. Given the astronauts' relatively small salaries, some thought the official reaction overblown; Irwin in particular was known to be a particularly religious person who would not knowingly do anything unethical. But others were appalled at the notion of any commercialization or profit on the astronauts' part. After all, it was taxpayer money and 400,000 laborers who had made Apollo possible, not primarily these three individuals.

It was a shame that any black mark accrued to *Apollo 15* in hindsight, for the crew were arguably the most energetic and geologically competent, and for sheer visual spectacle the Hadley site represented a dramatic pinnacle among all the surface explorations.

During the congressional investigations it came to light that *Apollo 14*, too, had engaged in commercial activity. The crew had carried 200 privately minted silver medals in their personal effects. These were later sold to the Franklin Mint, which had then made a nice profit melting and recasting them into 130,000 commercial commemorative coins "which had been on the moon."

The floodgates of inspection were now open: it further came to light that fifteen astronauts had sold 500 copies of each one's signature on blocks of stamps, and received $2,500 apiece, though five of the men had donated the entire proceeds to charity.

That one was an iffy situation, because this latter group had not transported anything commercial to and from the moon; they had merely sold their own signature, which presumably every human being has a right to do. Nonetheless, Deke Slayton and Al Shepard were upset and exasperated by the entire business, since they'd not gotten wind of any of it until it came out in the press.

Deke Slayton personally delivered lectures to the upcoming crews that nothing like this must ever again occur.

It didn't.

◯

# CHAPTER 23

# Busting the Moon's Speed Limit

*The moving moon went up the sky,*
*and nowhere did abide;*
*Softly she was going up,*
*And a star or two beside.*

—Samuel Taylor Coleridge, *The Rime of the*
*Ancient Mariner*

When *Apollo 16's* Saturn blasted off on April 16, 1972, it was the heaviest object ever to leave the planet Earth. Lying supine in the left seat was Commander John Young, the veteran of *Gemini 3* and *10* and *Apollo 10*, which circled the moon but did not land. Joining him was LM pilot Charles Duke—and CSM pilot Thomas Mattingly.

Everyone now knew that there would only be one more moon landing after this one, so a Command Module pilot such as Mattingly, who would merely circle the moon but never reach the surface, might seem to have landed a rotten job. The truth, however, was that he was glad to be there. NASA had recruited and trained far more astronauts than would ever fly. They were getting to be like mice.

Deke Slayton—who through novel medication had now overcome his heart arrhythmia and had just been returned to flight status after a decade in the high-status desk job as astronaut crew chief—announced publicly that NASA had three times more astronauts than they presently needed. Slayton had urged those who had not already been assigned to either the final Apollo *(17)* or the upcoming three-flight Skylab space station missions, to seriously consider looking for work elsewhere.

So a spot aboard an Apollo, even if it was merely to orbit the moon for three days, was still a highly coveted position, and nobody regarded it lightly.

*Apollo 16* was destined to be very successful but nonetheless trouble-prone. Numerous problems—with leaking helium bladders, warning lights, balky space suits, and especially the backup SPS engine motor—had to be dealt with and solved during the twelve days away from Earth. That latter problem was so serious, it nearly scrubbed the landing when the two craft, named *Orion* and *Casper*, had already separated in moon orbit.

The motor that physically steered the giant seven-foot-diameter SPS engine under the command module, showed oscillations. These could be stopped using the primary unit, but mission rules required that both systems be operational. Houston put a "hold" on the landing for a full six hours, while the two spacecraft circled the moon a dozen times within sight of each other, before endless frantic meetings with system contractors devised a way to make the backup work well enough if needed, to allow the men to get a "go" for the landing attempt. But it threw the entire schedule off.

On the ground after an uneventful landing, in the most southerly of all the sites, the glitches continued. As the many experiments of the ALSEP package were being laid out not far from the LM, with cables and wires splayed along the ground, Commander Young tripped over the ribbon wire and yanked the connections out of the heat flow electronics package.

Young: "Something happened here."

Duke: "What happened?"

Young: "Here is a line that pulled loose. Uh-oh. What is that? What line is that?"

Duke: "That's heat flow. You've pulled it off."

Despite frustrated efforts from Houston to find a way to reconnect the data cables, it was to no avail. The experiment was completely lost.

Later, the orange juice container inside one of the space suits ruptured, spilling sticky liquid into the suit. Later both men worked to try to get the helmet off, but it was firmly stuck, glued in place by the dried sugar. Only when water was deliberately spilled inside could they remove it.

But despite such glitches, the EVAs went splendidly.

With only two landings left, scientists were increasingly desperate to find volcanic material, or other rocks other than the simple breccias that seemed to be everywhere. Breccias are clumped-together surface powder, fused by the impacts of meteors. Thanks to the moon's endless bombardment over the course of aeons, the surface is littered with this type of rock, and they told geologists nothing about the moon's composition farther down. Volcanic material would do that, but the samples from the first four landing sites had uncovered none whatsoever. This was considered the consequence of having gone to safe flat areas, and *Apollo 16* and *17* were sent to more favorable sites with, scientists hoped, volcanic histories.

Actually, scientists really wanted *16* to go to Tycho, the enormous sixty-mile-wide crater whose radiating white rays splay dramatically across much of the moon's face; it can even be seen from Earth with ordinary binoculars. A meteor created Tycho, blasting out an impressive chunk of the crust of the lunar highlands and throwing strings of luminous material for more than a thousand miles. That entire region made scientists drool with longing.

"You will go to Tycho over my dead body," said the head of the Project Apollo office, astronaut Jim McDivitt, who used his veto power because he thought the area was too dangerously rugged. So this new site, the hilly terrain near the crater Descartes, would have to suffice. It looked like it might have had volcanism in its past, and was a promising geological site.

So now the men were here, and got busy. They spent more than twenty hours on the surface outside the LM. But by the time it was all over, not a single volcanic rock was found. Moreover, the site was less interesting, physically and visually, than the dramatic vistas of *Apollo 15*.

On the first drive, which gathered the largest single rock (twenty-five pounds) of the entire Apollo series, the men finished by demonstrating the LRV's capabilities to its designers watching from back on Earth. The scenes of this "moon buggy" making sharp turns and kicking up clouds of lunar regolith, and leaping into the air over bumps, became famous film fare in later years, and were popularly regarded as the astronauts' simply having a good time at taxpayer expense. In reality, the men had been asked to put the machine through its paces to test its capabilities "at the edges," though nothing of course prevented them from enjoying themselves at the same time.

Duke, a serious amateur photographer who had once filmed the Grand Prix, now took a movie of Young driving the rover. "Man, I'll tell you," he explained, "Indy's never seen a driver like this!"

When they came back in they were so covered with the clingy moon powder that, dusting off their suits, it filled the LM's air in a thick cloud. Young got a mouthful of it, and said later that it didn't taste bad. He didn't recall if he actually swallowed any, but he became possibly the only human being to have sampled moon in the culinary sense.

The second EVA lasted more than seven hours and took the men up the steep twenty-degree slope of Stone Mountain. They actually climbed halfway up the 1,400-foot-high mountain, a full 700 feet above the area where the LM sat. It was so steep, one astronaut had to hold the vehicle to keep it from sliding down whenever the other got out to gather samples. "What a view," Duke crowed, and later images (the men shot over 10,000 still photographs) showed that this was no exaggeration.

On the way down they "opened 'er up" and set the official lunar speed record of eleven miles per hour.

The final EVA, on their third day on the surface, lasted "just" five hours. The pair drove to the rim of the largest crater visited by any of the Apollo personnel. North Ray Crater was more than a half-mile wide, with an enormous boulder sitting on its rim. Some sixty-six feet tall, as much as a six-story apartment building, the men dubbed it the "house boulder," and it figured prominently in their photographs.

But North Crater offered one major surprise—an area of magnetism that was the strongest recorded by any astronauts in the entire Apollo program. The moon is magnetically dead, having no overall dipole field the way Earth and most other planets do. Rather, its surface does contain small spots of local magnetism, and this was the strongest inspected from the ground. Astronomers believe that a metallic body that crashed into the moon to form North Crater—probably an asteroid—is responsible for the odd magnetism.

Before Charlie Duke went back inside, he did something that other astronauts later wished they had thought of: he left a laminated photograph of himself, his wife, and his two children, ages five and seven, along with their signatures on the back. He took a photograph of that photo lying on the surface, and that remains to this day his fondest possession.

The three-and-a-half-day trip back to Earth passed uneventfully except for some definite crankiness noticed in the crew's demeanor. Following heart arrhythmias experienced by both men of *Apollo 15* while on the moon, Dr. Berry's medical team decided to increase potassium in the astronauts' lunar diet and also give them more rest periods. But they still were overworked, still had to deal with all the glitches, and, worst of all, had their entire schedule revamped because of the six-hour delay in getting to the surface. Added to the astronauts' always-present day-night disruption of the human circadian rhythm, it all proved once again that the moon-going experience was not easy psychologically or physically, and that it took a special kind of person to handle this kind of mission without cracking.

Perhaps all those "blank white paper" psychological tests had served some purpose after all.

# CHAPTER 24

# Final Footprint

*How sweet the moonlight sleeps upon this bank.*

—William Shakespeare,

*The Merchant of Venice*

A*pollo 17* was science's last chance at the moon. Of course, at the time, everyone thought that once the Apollo program was complete and the government paused for a few years to catch its breath, the United States would march onward to Mars, to moon colonies, to all manner of repeated manned exploration. Nobody in 1972 would have imagined that the twenty-first century would be well along before the United States would even once again *think* about returning men to other celestial bodies.

But for now, this was it. Vacationing families in Florida, suddenly waking up to the fact that Project Apollo was ending, grasped that this would be the final chance to watch a rocket depart for another world. It was also the only night launching of the entire Apollo program, and the crowds fully returned, lining the causeway as they had done for Armstrong and Aldrin two-and-a-half years earlier.

The lunar target for this final try was a six-mile-wide valley in the Taurus Mountains. For the first time, an actual professional geologist, Dr. Harrison (Jack) Schmidt, would be on the crew. This was the only time that a non–test pilot would walk the lunar surface, and he was a civilian to boot. Joining him would be Commander Gene Cernan. Orbiting around in the CSM would be Ron Evans.

As the men waited atop the thirty-six-story-tall rocket, there was a strange disconnect between the high-tech event about to unfold, and the mundane decay that had already consumed the aerospace community. Already the Apollo program's contractors had fired more than 300,000 workers, down from the 400,000-plus personnel of its heyday. In and around Cape Kennedy, houses were being put up for sale, real-estate prices were plummeting, banks were closing. The future of spaceflight was very much in doubt.

As an indicator of public apathy beyond the immediate Cape Kennedy area, none of the networks interrupted their programming to show the countdown to this final launch to the moon. Even PBS turned down NASA's request for live coverage. The scheduled blastoff in prime time didn't help either; the launch window began at 9:53 p.m., and no network executive was willing to cut into this top advertising revenue time. Humankind's final trip to the moon simply could not displace *All in the Family*.

Among the personnel at Cape Kennedy, pride was nonetheless high. They were determined that this final flight should not meet with tragedy due to anyone's apathy or slipshod work. Despite this care, the launch had glitches.

When the countdown had proceeded all the way to T minus 30 seconds, it abruptly froze—a first in NASA's history. What had transpired was quirky: at T minus 2 minutes, 47 seconds the automatic sequencer was supposed to pressurize the Saturn's third-stage (S-IVB) liquid oxygen tank, but failed to do so. Controllers monitoring all systems saw this fault and pressurized the tank manually, but a minute later the sequencer realized its own mistake (it had been caused by a faulty diode), did not register that humans had already compensated, and merely stopped any further activity.

The only thing that could be done now was to back the count up a full twenty-two minutes, a process that itself took forty minutes to accomplish, along with other necessary holdups. The bottom line was that the crew would be delayed reaching orbit and the crowd's party would be a late one. Finally at 12:38, a half hour after midnight, the countdown reached zero and the Saturn's five F-1s ignited.

Everyone who saw it was astounded. The night launch was nothing short of unbelievable. Night literally turned to day, and the glow could be seen all the way to South Carolina. The deep roar of the engines pounded the chest cavities of onlookers ten miles away—it was simply the loudest sound ever created by humans. *America*, the CSM, and *Challenger*, the LM, were on their way to the moon.

One minor oddity of the crew's outward leg happened near the end of the final day: Houston couldn't wake the sleeping men. First they simply didn't respond, and then it took a full hour to rouse them. Houston even debated transmitting a high-pitched oscillator signal as a makeshift alarm clock. Mission Control tried yet again.

Houston: "Hello *17*, hello *17*. How do you read us this morning?"
*Apollo 17*: "We're asleep."
Houston: "That's the understatement of the year."

When the time came, the two craft separated and *Challenger* began its twelve-minute burn down to the surface. Like the two missions before them, they dramatically swooped in just above mountains, and in this case descended into a valley beneath the level of tall hills on their right and left. Gene Cernan had come close to the moon exactly three years earlier on *Apollo 10*. Now he was allowed to keep going.

Cernan and Schmidt took over manual control when they were 300 feet up, watched dust obscure the terrain below them when the LM was sixty feet up, and when Cernan saw the blue contact lights come on, he killed the engine. The craft dropped in from five feet and touched down at two-and-a-half miles per hour, the LM's rear leg settling back into a small twelve-foot-diameter crater. They were just five degrees from vertical. Good enough.

Cernan was so psyched, he was practically hyperactive: "Okay Houston, the *Challenger* has landed. I shut down and we dropped, didn't we? Yes sir, but we is here. Man, is we here!"

Cernan now went through his checklist to make sure all was well, that they could remain: "I'll check everything again. It looks good—the manifold hasn't changed, the RCS hasn't changed, ascent water hasn't changed, the batteries haven't changed. Oh my golly—only *we* have changed!"

He had indeed. He had joined the small exclusive club of humans to have visited the moon.

The men had a very ambitious schedule—and this mission proved to be the most scientifically productive of all the landings. First priority, other than setting up the usual instruments plus several new ones, was searching for volcanic traces. Many scientists were still convinced that not all craters were caused by meteor impact, but that some had been created by long-ago volcanoes. In this site—a rugged wonderland that vaguely resembled the dream-scenery of *Apollo 15's* Hadley plain—several suspicious-looking cones had been sighted from previous orbital photos. If volcanoes had been here, then so should volcanic material.

On the second EVA, Schmidt and Cernan ecstatically thought they'd found it. There in the unrelenting gray of the lunar terrain were patches of orange. Surely this must be some mineral spewed from ancient volcanoes. Geologists back on Earth thought so, too—but when later analyzed, the samples were merely more regolith, colored by chemistry. Lunar volcanoes, it seemed, were no more than moon mirages.

Schmidt also set up a string of grenades that would be fired remotely after the astronauts were safely home. Flying a half mile through the sky, they'd detonate with a precisely known force, generated underground waves that would be picked up by the various seismometers this crew, and previous ones, had laid down at the six sites. Since acoustic waves travel differently through different materials, researchers on Earth hoped to learn the moon's underground composition.

All the scientific equipment left on the moon (except for that of the *Apollo 11* group, which failed in just a few weeks) continued to operate for years, thanks to a reliable power supply: plutonium. As were later unmanned probes to Jupiter and Saturn, the ALSEPS were powered by an RTG, a radioisotope thermoelectric generator that would convert the heat of decaying plutonium directly into electricity. Solar power could have been used by day, but the bitter two-week lunar nights, with temperatures down to 250 degrees Fahrenheit, made solar power untenable.

This mission would also see the first lunar auto-body repair, after Cernan swung a hammer and accidentally ripped off the thin right rear fender from the rover. Without it, a stream of lunar dust would continually strike the riders and cover them.

Dust was not a minor problem. Already the baby-fine powder was getting on the space suits, clogging equipment and cameras, and clinging electrostatically to nearly everything. It unavoidably found its way into the LM and, once in orbit, even sometimes managed to migrate into the Command Module to accompany them all the way home. They lived with it and breathed it: the moon literally became a part of all the Apollo astronauts' bodies.

So now, without a fender over the wheel, the spraying dust would reach impossible—and deadly—intensities. Each space suit's cooling system depended on the water in fine pipes that kept circulating within the material; this water was itself cooled by a special radiator on their backs that allowed a small amount of water to evaporate into the vacuum of space. The evaporation process is always accompanied by cooling, which is why nature makes humans perspire in the first place. The point was: if these radiators became clogged with fine dust, they would not work, and the occupant would overheat. For this reason, Apollo astronauts periodically dusted off their companions' suits.

So, riding the now-broken LRV with the missing fender would not merely be unpleasant and dirty; it could be lethal.

A fix was radioed in from astronaut John Young in Houston. Spending hours on the problem, he devised a jury-rigged fender and instructed Cernan and Schmidt how to construct it. They used thick laminated maps of the surface bent to form the correct curved shape, duct tape, and clamps removed from the LM's telescope. It worked, and *Apollo 17's* EVAs could continue.

When the men were finishing up their final EVA on the third day, the sun had risen 40-degrees high, roughly halfway up the sky, and the ground had heated up to 150 degrees Fahrenheit. It was time to go.

Cernan and Schmidt had broken all records. They had spent seventy-five hours on the moon, with twenty-two hours of that doing EVAs outside the LM. They had driven the longest distance—a total of twenty-one miles. In fact, they had managed to cover the entire valley in which they had landed, going six miles in one direction from the LM, and an equal distance in the other, from the mountains bordering the southeast valley floor to those of the northwest side.

The people of Earth saw little or none of it. Although the color TV images were the clearest yet, the networks and PBS carried almost no live broadcasts from the moon. Nor, unfortunately, had NASA built up excitement by finding some genuinely charismatic TV personality to generate any sense of interest and adventure. Like the government bureaucracy that it was, NASA seemed unable to pull off the theatrics that the event naturally invited.

It remained for those who cared about science or adventure to savor NASA's amazing achievement in the days, years, and decades to come. In this single *Apollo 17* odysscy alone, humans had brought back 243 pounds of moon specimens, and spent twelve-and-a-half days off the planet Earth.

Schmidt had climbed up the ladder and was inside, and Cernan made a nice little speech for the camera: "We leave as we came, and, God willing, as we shall return, with peace and hope for all mankind."

The video camera aboard the rover stood by, and recorded the noiseless blast of the ascent engine, and panned up to watch the LM climb and vanish into the black sky. Then it panned back down where the debris kicked up by the engine's thrust had just settled. The camera's controller back on Earth, seemingly at a loss as to what to do next, panned back and forth for a few minutes, showing an inanimate scene that would remain unchanged for a half billion years.

The batteries were good, the camera could operate for a full day longer. But there would be no further changes, nothing more of import, and anyway no one was watching. Without a further word of narration, it was simply switched off.

○

# CHAPTER 25

# Who Were the Men of Tranquillity?

> *War talk by men who have been in a war is always interesting; whereas moon talk by poets who have not been on the moon is likely to be dull.*
>
> —Mark Twain, *Life on the Mississippi*

A total of seventy-three men had successfully earned the official title of "astronaut" by the time the Apollo program had finished. All had to pass through the incredibly difficult selection process. All were smart, and courageous, and physically fit. But the biographies of seventy-three people would fill this entire volume, and induce sleep.

We could limit it to the thirty astronauts of the first three groups selected. Or the forty-three who flew aboard *some* mission between Mercury and Apollo Soyuz. Or just the twenty-seven who crewed one of the Apollo missions and at least ventured as far as orbiting the moon. But for our purposes, that's still too many. And anyway this would still unfairly eliminate the pathfinders who rode Gemini and Mercury, and those such as Deke Slayton who were fully qualified but for whom circumstance prevented them from flying Apollo.

In reality, each certified astronaut first served as a backup crewman, and trained exhaustively in that capacity for an upcoming mission that they only had a minuscule chance of actually flying. Then, by custom, they would skip the next two missions but become a prime crewman for the third one ahead.

This procedure worked out well, except that some were bumped, some were left marooned on Earth by the cancellation of the final three Apollos, and some simply came into the program too late to be anything but back-ups. There were even a few odd cases, such as the crew who accompanied Wally Schirra on the Earth-orbiting *Apollo 7* only to be effectively banned from future flights to the moon because of Schirra's "insubordination." The bottom line is that many eminently qualified astronauts never went to the moon.

Then there are those who served as CSM pilot, whose job was to remain in lunar orbit while his colleagues voyaged down to the surface. Six people had that experience. Count them in or out of the "went to the moon" biographies?

At the end of the day, one has to draw an arbitrary line somewhere, simply to limit the bios to a manageable number. We'll make it an even dozen, and restrict it to the men who actually walked—or hopped—on the lunar surface. These were not better men than the others. They were just luckier.

In order of their contact with the moon:

# Neil A. Armstrong

Neil Alden Armstrong was born on August 5, 1930—the year of nativity of nearly half the astronauts. Born on his grandfather's farm in Wapakoneta in northwestern Ohio, he, like nearly all the other moonwalkers, was the oldest child. Neil had a sister and a brother.

Armstrong's biography looked storybook-promising from the start. As a child he loved aeronautics and built countless model airplanes. At the age of fifteen he started taking flying lessons with money he had earned doing odd jobs, and qualified for his student pilot's license at the legally minimum age of sixteen.

After graduating from Blume High School in Wapakoneta, he was accepted at Purdue University on a navy ROTC scholarship, and started studying aeronautical engineering. But his college years were cut short by the Korean War; he was called up for active duty, given flight training at Pensacola, Florida, and sent to Korea. He ended up flying seventy-eight missions from the aircraft carrier USS *Essex*, and was awarded three medals.

When the war was over, Armstrong returned to Purdue to complete his degree, which he attained in 1955. His first job was as a test pilot with NACA, the National Advisory Committee for Aeronautics, the forerunner to NASA. As a civilian, he conducted flight tests at the Lewis Flight Propulsion Laboratory in Cleveland, and later relocated to the famous Edwards Air Force Base in California, where he flew a half dozen experimental planes including the renowned X-15.

When NASA announced that they were looking for astronaut candidates, Armstrong put in his name, and was selected to the second group, the Next Nine. Although he had started too late for the Mercury program,

Armstrong rode into space aboard *Gemini 8*. (His adventures on that flight are recounted elsewhere in this volume).

Along the way, Armstrong met his wife, Jan, and managed to find time to start a family, eventually becoming the father to two sons.

After gaining the singular fame of being the first man on the moon, aboard *Apollo 11*, Armstrong eschewed publicity and granted very few interviews over the subsequent years. He could, like some other astronauts, have parlayed his celebrity into riches, but chose instead to accept a quiet job of teaching aeronautical engineering at the University of Cincinnati, although he did for a few years take the chairmanship of a small computer systems company.

To this day he avoids the limelight, refusing the steady stream of requests for interviews and TV appearances. He prefers to live quietly in Ohio, choosing to spend time with family and a small number of close friends.

While other astronauts have been vocal about how going to the moon either changed them or did *not* change them, Armstrong remained private. In 1970, *Life* magazine published *First on the Moon*, which although it credits the crew of *Apollo 11* as "authors," is made up of long passages of quotations from the astronaut's interviews with the magazine. He was asked during a television interview some years after the mission what the greatest moment was, upon landing on the moon. Stepping out upon the surface? Uttering his famous "One small step" speech? Neither. He said the highlight for him came just after the LM's engine had been stopped and the dust had settled. He and Aldrin looked at each other and silently shook hands. "*That* was the best moment," he said.

Bad interview subject—fascinating person.

# Edwin E. Aldrin Jr.

Buzz Aldrin was born in Montclair, New Jersey, on January 20, 1930. He was the only son of U.S. Army Colonel Edwin Aldrin and his wife, Marion. Edwin's father was a renowned pilot who personally knew or worked with such early aviation legends as Billy Mitchell, Orville Wright, and Charles Lindbergh.

His grandfather on his mother's side, amazingly enough, was an army chaplain with the odd and prophetic name of "Moon."

Edwin acquired the nickname "Buzz "from his older sister when she was a toddler and called him "buzzer" because she couldn't pronounce the word "brother." He liked it so much he adopted it throughout his life, and finally legally changed his first name a decade after he returned from the moon.

After graduating from Montclair High School, and showing aptitude for math and for various sports, especially pole vaulting, he attended West Point and graduated with honors in 1951.

Selecting the air force for his commission, he attended basic flight school in Barstow, Florida, and went onto jet fighter training in Texas, where he won his wings in 1952.

Aldrin was sent to Korea, and flew the F-86 on more than sixty-six combat missions. After the war, he went through a series of air force jobs that included gunnery school instructor in Nevada, before joining the new Air Force Academy as a flight instructor.

After getting married, Aldrin accepted an overseas post with his family in Germany in 1956, and became flight commander of a fighter wing. Three years later, he was back in the United States to attend MIT, where he attained a doctorate in astronautics. When astronaut candidates were announced, he submitted an application, and was accepted to the third group (the Fourteen) in 1963.

Aldrin's smarts, skill, and physical fitness got him into position where he could walk on the moon. But sheer good fortune landed him the coveted spot aboard the earliest team to touch down.

First, the normal rotation would have made him eligible to fly on the thirteenth Gemini mission, except there was no such flight: the final one was *Gemini 12*. Had he never flown on Gemini, the untested space-rookie would not have been given a flight aboard one of the all-important first Apollos. But fate stepped in with the sudden deaths of Charlie Bassett and Ed See, the crew of *Apollo 9*, who perished in a plane crash. This meant that their backups became the prime crew for their *Apollo 9* flight, and all other crews were moved up one rotation, which gave the final *Gemini 12* slot to Aldrin.

Aldrin turned out to be the star of that flight when the primary radar failed. For him—and NASA—this electronics crash proved to be the best possible thing that could have happened. That November 1966 mission was slated to test spacewalking and rendezvousing with a target Agena vehicle. With the radar failure, the rendezvous proved problematical; however, Aldrin's specialty was orbital science and maneuvers. He believed that astronauts using only special star charts, a sextant, and their own calculations should be able to determine precisely how to catch up with any other vehicle in a reasonable orbit. When Aldrin did so manually, without the aid of computers, his reputation as "Doctor Rendezvous" was assured.

Next, his long, dedicated, self-imposed study and water tank training in performing EVAs let him show everyone on Earth how spacewalking should be done. Others before him often had bad experiences where they'd tumble out of control, were unable to perform the tasks outside the module, work to exhaustion, and even have to come back in early. Aldrin, by contrast, performed his six-hour EVA flawlessly: the long, focused self-discipline and training had paid off. When the flight was over, he had joined the rarefied ranks of the "best of the best."

As for being on that first landing flight of *Apollo 11*, Aldrin's "number" just happened to come up right. It could have gone otherwise, but it didn't, for there had been an excellent chance that either *Apollo 10* or *12* would have been the first to touch down, and even the *Apollo 11* job had reportedly been initially offered to another astronaut, who had turned it down.

Later years were very mixed for Aldrin. Returning from the ordered world of the military and then the astronaut corps, to a life of a fame that no one could live up to, drove him into a period of deep depression and alcoholism. Eventually, Aldrin checked himself into a mental hospital.

"[As an astronaut] everything I did was very structured," he told this author years later, "then afterward things were really unstructured. That's when you can inherit tendencies that existed in your parents and your family." Fortunately he forged through this era and came out the other side in good shape, married a second, and then a third time. This marriage is still going strong, and now Aldrin is a content grandparent living in Southern California.

"Youth is great, but this is the most satisfying time of my life. I'm far more competent and creative than I was then."

Reflecting on the many twists his life has taken, he says, "I'm not a crusader. I'm a survivor."

He doesn't think that being on the moon per se, changed him or influenced his personality and behavior. Rather, it was having to deal with being a celebrity, though he modestly doesn't use that word to describe himself. As for passions, he remains an active scuba diver and skier, a sport he had never tackled until the age of fifty.

Aldrin's articulate, extroverted nature manifests itself: As the polar opposite of Armstrong, he delivers lectures and appears on TV shows. He's friendly and quick to discuss space matters, and is still very involved with innovative orbital procedures—now directed at the goal of reaching Mars. But he has little patience with know-nothings. Buzz punched a heckler in the 1990s who had followed him around insisting that he was a fake and that humans never really did land on another celestial body. By all accounts, he very nearly sent *that* human to the moon.

# Charles Conrad Jr.

As Commander of *Apollo 12*, Conrad was the third person to step foot on the moon. His life started as Charles Conrad Jr., on June 2, 1930, in Philadelphia. His mom, however, kept calling him "Pete" as a nickname, since that's what she wanted to name him in the first place, and eventually he adopted it permanently.

His father was an aviation buff as well as a World War I hot-air balloonist. Pete would build mock-up airplane cockpits in his room as a child, and took flying lessons in his early teens. A child of privilege, he attended private schools in Pennsylvania and a boarding school in New York before attending Princeton to study aeronautical engineering.

In rapid order he earned his undergraduate degree, got married, and joined the U.S. Navy to become an aviator, and eventually earned the right to fly jet fighters. He loved aviation, and enrolled in the U.S. Navy Test Pilot School in Maryland, where he served as an instructor from 1959 to 1961. When NASA announced that they were looking for astronaut candidates, he put in his application. Although he failed to make the first cut for the Original Seven, he applied again and was accepted to the second group, the Next Nine, in September 1962.

Conrad got his first taste of the vacuum of space in August 1965, aboard *Gemini 5*. This was the longest-to-date trip into space, an eight-day struggle in a cramped unpleasant space, but it proved the reliability of the electricity-generating fuel cells that would be needed to go to the moon, and proved that humans could survive that long in a gravity-free environment.

Conrad flew yet again on *Gemini 11* a year later, and performed a spacewalk. Now a seasoned astronaut, he was selected to *Apollo 12*, the second manned lunar landing, which left Earth on November 14, 1969.

The wild adventures of *Apollo 12* in November 1969, including being struck by lightning shortly after liftoff, are recounted elsewhere in this book. Their pinpoint landing, just 200 yards from the robot *Surveyor* spacecraft, demonstrated the capabilities of the newly refined lunar navigation procedures. As Conrad stepped down—as one of the shortest of the astronauts at five foot, seven inches—he had to leap off the final rung of the ladder, and uttered his famous words: "That may have been one small step for Neil, but it's a long one for me."

Pete Conrad's flying career was still not over after he returned from the moon. He was chosen to command the premier Skylab mission in June 1973, a twenty-eight-day stint aboard America's first orbiting space station. Problems with the jammed solar panels first necessitated complicated repairs, and Conrad and fellow astronaut Joe Kerwin labored outside until they successfully got the ship working again, and then settled down to their record-breaking stay in Earth orbit.

He then left the space program and retired from the navy with the rank of captain, beginning a series of management positions first with a cable company, then a long-term relationship with the aerospace company McDonnell Douglas.

When reflecting on his four trips into space—totaling about a month and a half off planet Earth—and his two spacewalks, and his hops around the moon's surface, Conrad has been repeatedly asked if all this has fundamentally changed his personality. His answer has been consistent and direct: "Not at all."

On July 8, 1999, at the age of sixty-nine, Conrad was motorcycling with friends when his Harley-Davidson motorcycle failed to negotiate a turn on Highway 150 near Ojai, California. He ran off the road and crashed, and died in a nearby hospital later that day.

# Alan L. Bean

Alan LaVern Bean was born on March 15, 1932, in Wheeler, Texas. His mother ran an ice cream parlor and a grocery, and his father was a government flood control expert. Like nearly all the astronauts, he was the oldest child. Because of being frequently transferred to where he was needed, Bean's father (along with his mother and a younger sister) were forced to move several times before finally settling down in Fort Worth.

Alan graduated from Paschal High School in Fort Worth. He dreamed of flying, so he joined the Naval Air Reserve, and enrolled at the University of Texas on a navy ROTC scholarship. He was athletically talented and excelled as a college wrestler and gymnast, and earned his undergraduate degree in aeronautical engineering in 1955.

The navy sent him for basic flight school in Pensacola, Florida, followed by advanced training that earned him his wings in 1956. After serving in a jet fighter squadron for four years, he entered the U.S. Navy Test Pilot School in Maryland, and remained there as a military test pilot while also taking night classes at St. Mary's college to study—of all things—art.

Bean applied when the call went out for astronaut candidates, but failed to make the final selection. He applied again the next year, and this time was successful, becoming a member of the third group, the Fourteen.

Unfortunately, and to his bitter disappointment, there were just not enough Gemini flights for him to fit in the rotation, and it looked like Apollo was going to pass him by, too. Superiors suggested he train for the Skylab missions, which would follow Apollo in 1973. But—as happened with several other astronauts—Lady Luck now made her entrance. When astronaut C. C. Williams was killed in a plane crash, a place aboard *Apollo 12* suddenly became vacant, and Bean's friend Pete Conrad, who was commanding the mission, recommended Bean to his superiors, who approved it. Bean would be going to the moon after all. His exploits there are documented elsewhere in this book.

Perhaps the greatest astronaut demonstration of *que sera sera* was exhibited by this *Apollo 12* crew once they were in Earth orbit, after knowing that their parachute system might or might not work after having been hit by lightning on the way up.

"We had to leave it to Houston to worry about that," Bean recalled years later.

After *Apollo 12*, Bean was selected to head the crew aboard the second Skylab mission, a record-breaking fifty-nine days in space that began in July 1973. When this was over, Bean and his two fellow comrades had suddenly usurped the total time-in-space record from all the other veterans.

Now, after Skylab and Apollo Soyuz, there would be a long hiatus of U.S. manned space effort while the new returnable Space Shuttle was being developed. Bean stayed with NASA through the long years of development as chief of the Astronaut Operations and Training Group. After many delays, the Shuttle was finally ready to fly in 1981, and Bean would have been one of the early crew to fly it. But that very year, 1981, he surprised everyone by quitting the astronaut program to start a new career as a professional artist.

His emphasis was to be the recollections, sights, and experiences of being on the moon, told in the way only a painter could tell it. Along the way, Bean divorced his first wife (they had two children, a boy and a girl) and married his new wife, Leslie.

Did going to the moon change his life?

"I feel more self-satisfied and happy," he said many years later when introspecting about his years as an astronaut. "I feel lucky a lot, and I feel blessed." Then he added, somewhat contradictorily, "I feel like somebody up there loves me a lot even though I don't believe in that kind of thing."

These days he *still* characterizes his profession as "full-time artist."

# Alan B. Shepard Jr.

Alan Shepard was born in East Derry, New Hampshire, on November 18, 1923. An eighth-generation New Englander, he attended a one-room elementary school, then graduated from the Pinkerton Academy in Derry in 1940. He became a midshipman at the U.S. Naval Academy in Annapolis Maryland, and served two years at sea in World War II aboard the destroyer USS *Cogswell*. Hoping for a switch to aviation, he completed flight training in Pensacola in 1947, earning his wings and becoming a squadron fighter pilot. Shepard then served repeated tours of duty in Europe as a carrier-based fighter pilot.

Shepard began training to become a test pilot in 1950 at Patuxent, Maryland, and graduated to enter his new field of flight testing and high-altitude research.

He didn't apply to be an astronaut candidate: he was one of 110 military test pilots personally invited to try out for the new job in 1959, and was chosen as one of the Mercury Seven. As such, he gained instant celebrity, with the famous *Life* magazine tell-your-personal-story contract giving him more money in a lump sum than he had ever had in his life.

As things turned out, Shepard was chosen to be the first American astronaut to ride into space, the now-famous fifteen-minute suborbital flight aboard a Mercury capsule he named *Freedom 7*. It resulted in wild ticker-tape parades and a visit to the White House. Usually, one gets fifteen minutes of fame. In this case it was a fifteen-minute flight that was to bring a lifetime of name recognition.

Shepard, a savvy businessman and investor, parlayed everything he earned, and the universal trust in his name, into a business empire of success and wealth—building for himself and his new wife a mansion in Texas.

When the Apollo program followed, he was given the commander's job on *Apollo 13*. But—and here is where destiny once again played its epic role—he felt he needed more time to prepare and requested a switch to the following flight instead. The *Apollo 14* crew were happy to move themselves up one flight and the switch was agreed upon by all. Smart dumb luck: *Apollo 13* never made it to the moon's surface, and instead Shepard now had command of *Apollo 14*, along with space rookies Ed Mitchell and Stu Roosa.

Bewilderingly, NASA was now handing command of this first flight after the ruined one to a trio that had only fifteen minutes of in-space experience between them. This, perhaps more than anything else, showed the faith they had in this Original Seven astronaut.

Did going to the moon change him? "Not really," he has answered, and indeed insisted that it didn't really change any of the other men fundamentally, so far as he could see.

After *Apollo 14*, Shepard returned to his venture-capital business full time, where by most accounts he became one of the wealthiest of the Apollo astronauts. He and his wife, Louise, lived near Houston, where they raised two daughters.

In later life, Shepard was stricken with leukemia, and finally succumbed to that cancer on July 21, 1998.

# Edgar D. Mitchell

Edgar Dean Mitchell was born on September 17, 1930, in Hereford, Texas, near the New Mexico border, the oldest of three children. His family moved to Artesia, New Mexico, in 1935, and he lived there afterward for much of his life.

In this farmboy and cowboy region, it might be hard to see how Mitchell's interests took such a turn toward ESP, UFOs, and psychic research, until one realizes that he went to elementary school in Roswell, later to become the UFO mecca of the Western world.

He was fascinated by aviation at an early age and earned flying lessons by washing and scrubbing planes at the local airport—the classic fairy-tale pilot's childhood. He soloed at the age of fourteen, although he had to wait until sixteen, the legal minimum age, to obtain his pilot's license.

After graduating from Artesia High School, he was accepted to the Carnegie Institute of Technology, where he earned his undergraduate degree in 1952. Afterward he joined the navy with the hopes of becoming a naval aviator, and in 1953 received his commission after completing training in Newport, Rhode Island.

He earned his wings in 1954, completing advanced flight training in Kansas, after which he was sent to Okinawa. He reenlisted, switched to carrier flying, and flew attack jets in 1957 and 1958.

His next career move was to switch yet again, this time to a naval air development unit where he qualified as a test pilot. At this same time, he pursued postgraduate studies, and finally earned a doctorate in aeronautics and astronautics from MIT.

Armed with these prestigious credentials, he was made chief of project management for the Manned Orbital Laboratory, which later became incorporated into the new entity of NASA. Then in 1964 he became an instructor at the U.S. Air Force Test Pilot School at Edwards Air Force Base in California. It was at this point that he applied to be an astronaut candidate, and was accepted in the fifth group, along with eighteen others.

He became an astronaut far too late to fly a Gemini mission, but was selected for *Apollo 13*. As luck would have it, his commander, Alan Shepard, felt the team needed more time to prepare, and suggested switching for the next mission. The *Apollo 14* crew were amenable to this, and this is why that crew ended up on the near-disastrous *Apollo 13*, while Mitchell went merrily to the moon on *14*, on January 31, 1971.

Mitchell is perhaps best known on that mission for doing something that did not come to light until much later: performing a series of ESP experiments with four subjects back on Earth. (He claimed reasonable success, but it must be noted that every other scientist who looked at the results said it *disproved* ESP.)

This walk on the moon was to be Mitchell's only spaceflight. He retired from the navy and left NASA the next year, and devoted himself to his passion: psychic research. He founded, near San Francisco, the Institute of Noetic Sciences, dedicated to exploring the mind and consciousness, including such phenomena as ESP.

His interest in world religions, cosmology, and philosophy was later expressed by him as an ongoing fascination with "how it all began, where we came from and how we got here." And "the fundamental questions that neither religion nor science have ever fully answered."

All told, Mitchell's rejection of formal religion, and his later criticisms of NASA's priorities, didn't by themselves earn him raised eyebrows; rather, his endorsement of people who claimed supernormal powers such as Uri Geller (the "spoon-bender") and others who were widely regarded as charlatans or nuts, put Mitchell in a separate category so far as the other astronauts were concerned.

Mitchell certainly became the most thoughtful, if "out there," of the men who walked on the moon, with an unending capacity to examine the deepest philosophical issues, even if it sometimes meant embracing such questionable notions as the Hundredth Monkey Phenomenon. (He explained: "Once a successful event at any level of nature takes place, that information is instantly known to other members of the species.")

At any rate, as to the question about whether being in space or going to the moon had changed him, Mitchell says that the trip into space gave him a life change, a switch in direction: "Your reality shifts from Earth-center to Cosmic-center. You no longer see things in terms of Earth-centered reality."

Mitchell has two daughters from his first marriage. He married again, and currently lives in Palm Beach, Florida.

# David R. Scott

Dave Scott was born June 6, 1932, in San Antonio, Texas, the only son of Brigadier General Tom Scott. He graduated from Western High School in Washington, D.C., and completed a bachelor of science degree, from West Point, where he was fifth in his class of nearly 700.

Scott then went on to learn basic flight training at Webb Air Force Base in Texas in 1955, served in the Thirty-second Tactical Fighter Squadron in the Netherlands for four years in the late fifties, and then returned to the United States where he earned a master's degree in aeronautics and astronautics from MIT. After this he enrolled in the U.S. Air Force Test Pilot School, and later the Aerospace Research Pilots School.

Scott flew with Neil Armstrong on the historic *Gemini 8* mission, where a stuck thruster caused the ship to spin wildly.

Although their primary mission of docking was a success, they had to return to Earth prematurely. Scott next crewed aboard the Earth-orbiting

*Apollo 9*, then was given his own command in *Apollo 15*, which explored the spectacular lunar region around Hadley Rille.

After retiring from NASA, and leaving the air force with the rank of colonel, Scott became increasingly involved with aerospace business ventures, particularly the facilitation of private satellites.

He lives in California with his wife, and has two children.

# James B. Irwin

James Benson Irwin was born on March 17, 1938, in Pittsburgh, Pennsylvania. Like nearly all the astronauts in this biography, he was the oldest child, with two siblings. His parents, James and Elsa, moved to Orlando, Florida, then Roseburg, Oregon, and finally Salt Lake City, Utah, where he graduated from East High School in 1947. After graduation, he attended the U.S. Naval Academy in Annapolis, and then transferred to the air force after graduation, attending flight school in Texas.

By his own account, he did not like flying at first, and even contemplated quitting before developing a love of being in the air. He then went to the University of Michigan to receive a master's degree in aeronautics in 1957, and entered the U.S. Air Force Test Pilot School at Edward Air Force Base in California in 1960.

He applied when astronaut candidates were announced, but failed to make the cut in the third and fourth selections. Finally in 1966 he was chosen as one the Nineteen, the fifth group.

His sole flight, aboard *Apollo 15*, left Earth on July 26, 1971.

Of all the astronauts, Irwin reports the most pivotal, overwhelming personal transformation from being on the moon. Although he had "accepted Jesus" and was a fervently religious person at the age of eleven, he said he experienced on the moon "the presence of God. It has remade my faith. I had become a skeptic and I know I had lost the feeling of His nearness. On the moon the total picture of the power of God and His Son Jesus Christ became abundantly clear to me."

In a great twist of irony, Irwin's deep religious ideals did not shield him from the smear that the "stamp episode" threw upon the entire *Apollo 15* crew, even though Irwin himself had returned the money before the story broke. Nonetheless, in light of that black mark, and dim prospects of ever flying in space again, Irwin left NASA and retired from the air force the next year. Since then, he has pursued a ministry, speaking extensively about his religious experiences.

He and others created High Flight, a religious organization, and Irwin made several religious trips to the Middle East. Like Job, however, he found himself plagued by physical misfortunes: he was seriously injured by a falling rock on the slopes of Mount Ararat while searching for remnants of Noah's ark, and suffered a serious heart attack in the 1980s.

Jim Irwin lived for many years in Colorado Springs, Colorado, with his wife, Mary, and their five children, but suffered a final heart attack on August 8, 1991. This most zealously religious ex-moonwalker, the youngest of the men to walk on the moon, was dead at age fifty-three.

# John W. Young

John Watts Young arrived on this planet on September 24, 1930, the same birth year as five of the twelve who walked on the moon. Though born in San Francisco, his family moved to Orlando, Florida, when he was two. He was an honor student and track star at Orlando High; after graduation, he was accepted to Georgia Tech; he received his undergraduate degree in aeronautical engineering with honors in 1952.

After college, he joined the navy and after a year was accepted to flight school where he won his wings, and spent the mid-1950s as a fighter pilot. He then entered the U.S. Navy Test Pilot School, and after completion in 1959 became a program manager, testing weapons systems. He was a skilled jet pilot, and, commanding Phantom jets, broke two records in the "time to reach altitude" category.

He applied to become an astronaut candidate and was selected in the second round of nine, in 1962. He wound up being the first of this group to make it into space, and the first man outside the Original Seven to reach orbit, aboard *Gemini 3* paired with Gus Grissom in March 1965. This was the first time that a spacecraft crew changed its own orbit to a new one—a prerequisite for going to the moon.

Young's next flight was as a commander: *Gemini 10*, a three-day flight in July 1966, accompanied by copilot Michael Collins. It successfully and repeatedly docked with target vehicles.

Apollo brought Young not one but two trips to the moon. The first, *Apollo 10*, was the famous tantalizing shakedown excursion when Young not only orbited the moon, but had his LM detach and venture to within just nine miles of the surface. Then, three years later, aboard *Apollo 16*, Young, as Commander, accompanied by Charlie Duke as LM pilot, made it all the way—overcoming mechanical problems outlined in this volume's chapter dedicated to that mission.

After Apollo ended, Young, unlike most of the other moonwalkers, remained with NASA in preparation for the upcoming Shuttle program, and replaced Deke Slayton as chief of the Astronaut Office in 1974.

Young, together with copilot Bob Crippen, were the crew of the very first Space Shuttle. When they flew *Columbia* into space for its inaugural trial and landed it two days later, they'd become the first people ever to bring a craft back to Earth to a landing on a runway. Two years later, in November 1983, Young flew *Columbia* again, this time on a ten-day scientific odyssey employing its onboard Spacelab module.

Although Young does not hold the record time in space, he holds the record number of flights: two in Gemini, two aboard Apollo, and two aboard the Space Shuttle—the only astronaut to fly in all three programs.

John Young—even as he approached sixty—remained an active astronaut. He was the only lifelong career astronaut among the men who walked on the moon.

Today he still lives in Houston, with his second wife, Suzy. He has two children from a previous marriage.

# Charles M. Duke Jr.

Charles Moss "Charlie" Duke Jr. was born on October 3, 1935, in Charlotte, North Carolina. He graduated valedictorian from the Admiral Farragut Academy in St. Petersburg, Florida, in 1953, and was accepted to the U.S. Naval Academy in Annapolis, Maryland, where he graduated with a bachelor of science in 1957.

Transferring to the air force ("because I got seasick"), he enrolled in flight school and earned his wings in 1959. After serving with a fighter squadron in Germany for three years, he returned to the States and enrolled in MIT, where he earned a master's in aeronautics and astronautics in 1964, and enrolled in the U.S. Air Force Test Pilot School at Edwards Air Force Base in California. After completing the program, he remained as an instructor; while there he applied for the astronaut corps and was selected as a member of the fifth group, the Nineteen, in 1966.

No other member of the Nineteen walked on the moon; there were ample veterans in line from the previous four classes. But Duke joined the lowly support crew for the early Apollos, gained attention for his quiet, easy competence, and, at the request of several astronauts themselves, was assigned to be CapCom—the astronaut at Mission Control who communicates with the ones in space. When Armstrong and Aldrin were landing on the moon and we hear their voices recorded in history, it is Duke's voice that is also heard—saying things like, "Roger, *Tranquillity*, we copy you on the ground."

Duke got his moon-chance by training as a backup for the ill-fated *Apollo 13*, which made him a member of the prime crew of *Apollo 16*, that penultimate lunar voyage that blasted off in April 1972.

This was his only spaceflight. He retired from NASA in 1975, and after serving in the air force reserves, left with the rank of brigadier major a decade later. In between, he started two small companies, and served on the board of several others. His biggest life-altering event, however, happened in 1978.

Although he said that he did not sense God's presence on the moon, he experienced a life-altering awakening six years later. "I began to read the Bible," he explained afterwards. "I began to bring my life in line with Scripture. Now I try to be content in everything I am and where I am."

Duke started traveling the world, giving talks and testifying about the power of Christ, and how he, Duke, had committed his life to Him.

He makes his home in Texas with his wife, Dorothy. They have two children.

# Eugene A. Cernan

Gene Cernan was born on March 14, 1934, in Chicago. He graduated from Proviso Township High School in Maywood, Illinois, and went on to Purdue University, where he earned his BS in electrical engineering, and later a master's in aeronautical engineering from the U.S. Navy's postgraduate school in Monterey, California.

After Purdue, he joined the ROTC and entered flight training, and flew in fighter squadrons based in California. When astronaut applications were announced, the navy itself submitted Cernan's name, and he was subsequently selected in the third group in 1963.

Cernan first went into space aboard *Gemini 9*, joining Tom Stafford on a three-day flight in June 1966. During this mission he became the second American to walk in space, performing a two-hour EVA. Of all the Gemini and Apollo splashdowns, this flight ended up landing the closest to its intended point in the sea—less than a half mile of dead-on. (Cernan, an energetic, take-charge kind of personality, readily admits to feeling a competitive drive to do things better than anyone else had ever done it.)

Cernan then was assigned to the crew of *Apollo 10* in May 1969, which circled the moon and ventured just nine miles from the surface.

On his third trip to space he was Commander of *Apollo 17*, the final visit to the moon. As such, he became the last human to leave footprints in the lunar soil; before reentering the LM he made the "We'll be back" speech that has yet to be fulfilled.

Cernan left NASA a few years later to enter the business world, where he has been highly successful. He started his own company in 1981, became executive vice president of Coral Petroleum, and most recently became chairman of Johnson Engineering Corporation, which is a major NASA supplier of flight crew systems. In short, he has used his energy and fame to make a lot of money.

Asked many times in subsequent years, he replies that the moon experience was *not* life-altering by itself.

His hobbies continue to be flying, hunting, fishing, and competitive sports. He lives in Houston with his wife, and has a daughter from a previous marriage.

# Harrison H. Schmitt

Harrison Hagan Schmitt, called "Jack," was born July 3, 1935, in Santa Rita, New Mexico, the son of a well-known mining geologist. After graduating from Western High School in Silver City, New Mexico, he attended the California Institute of Technology, where he received his bachelor of science. He then studied at the University of Oslo in Norway, and then went on to Harvard where he earned his doctorate in geology in 1964.

When NASA announced that it was looking for scientist-astronauts, Schmitt applied, and was selected—in this special category—in June 1965.

As such, Schmitt's background was wholly different from all of the other men who walked the moon's surface. He had never even piloted a plane, let alone been a test pilot like the others, and now—at age thirty—was being sent by NASA to basic flight training school.

He reported that while he found visual flying not too difficult, he was ten years behind everyone else when it came to learning instrument skills. He was assigned to be a backup crewman on *Apollo 15*, which would have made him a prime crewman on *Apollo 18*. But *18, 19,* and *20* were canceled, and NASA made the decision—under intense prodding by the National Academy of Sciences—that they couldn't end the Apollo program without sending a single scientist to the moon. Schmitt was it.

This meant breaking up the scheduled crew for *Apollo 17* (splitting up already-trained crews was very rarely done) and pulling off one of the crew members who had worked and prepared so hard for it, and whose turn it was. The unfortunate victim was Joe Engle, who remained bitter for years afterward. (His subsequent flights on the Shuttle did indeed get him into space many years later, but nobody doubts that it was inadequate compensation for not walking on the moon.)

At any rate, a number of other astronauts quietly resented Schmitt as well: he was the only moonwalker who hadn't "paid his dues" in performing any military service whatsoever. But everyone accepted the reasoning behind the decision, that a trained geologist could accomplish more than the relatively short science training that the other astronauts might receive.

After *Apollo 17*, Schmitt left NASA in 1975 and made a successful run for the U.S. Senate on the Republican ticket, and served from 1977 to 1982.

A long-time bachelor, Schmitt finally married in his fifties, and took up the lecture circuit. He still makes a fine living writing, speaking, consulting, and occasionally teaching for the University of Wisconsin at Madison. He is also a member of several corporate boards.

○

# The Lunar Book
# of Revelations

*And there is nothing left remarkable*
*Beneath the visiting moon.*
—William Shakespeare, *Antony and Cleopatra*

Three American lives. Billions of dollars. The dedication of hundreds of thousands of professionals, some of whom labored on nothing else but the Apollo. What did we get in return?

It's hard to place a value on the national pride and prestige that going to the moon accrued. Or the benefit to the soul of such cutting-edge adventure and exploration. No one doubts that we'd all be poorer if men hadn't kicked up the dust on Tranquillity.

But, sticking solely with the science aspect of Apollo, what did we learn? What do we know now that we didn't before? How important is this knowledge? In crass materialistic terms, was it worth $24 billion?

Before we get to the specifics, it should be candidly noted that space science is not generally a very practical branch of knowledge.

Granted, the *engineering* needed for the moon missions brought innovations that found their way into industry and everyday life—practical things such as Velcro and Teflon, minor conveniences such as Tang breakfast drink, and computer advances that would have occurred in any case, but perhaps taken a bit longer. However, if one were to be honest about it, outside the realm of creative engineering, the field of astronomy is inspiring precisely because it's truly "pure" knowledge, knowledge without a practical purpose.

The public and a wary Congress, looking at taxes and funding, are not often sympathetic to purposeless science, so NASA has always tried to couch upcoming projects in terms of human or terrestrial benefits. The goals of the *Voyager* and *Galileo* probes to Jupiter, which cost a billion dollars each, were often marketed with the mantra: Exploring Jupiter will teach us more about Earth.

That's a stretch. In truth, we were learning about Jupiter because it's an amazing, fascinating place, well worth exploring in its own right. The practical benefit for earthlings in sending probes to Jupiter was between negligible and zero. Similarly, the Apollo missions were to learn about the moon, our nearest neighbor. There was admittedly some small earthly advantage in seeing how chemistry and geology work on another world. But the overwhelming reason to go was to answer ancient questions about the moon, period. There was no chance that the information would ever increase Earth's food supply, solve our energy problems, make our homes more secure, increase our life expectancies, or give us any other significant practical benefit.

In other words, politics and prestige aside, Apollo involved discovery, adventure, and science of the noblest sort—the kind done for the sheer human yearning to learn and to know. Some people are inspired by such an intent, some appalled.

Now, more than thirty-five years later, it's natural to ask: "What did Apollo teach us about the moon?" What's interesting is that very few educated people can offer any sort of answer. Even most high school physics teachers would be hard pressed to offer a single discovery produced by the dozen men who poked, prodded, and drilled their way into six different sites on the moon's surface.

Why the Apollo science harvest is not more widely disseminated is perhaps a mystery. As we've said in earlier chapters, NASA was probably remiss in not having, in its budget, a personable science speaker along with a skilled production crew, whose sole job was to translate the discoveries into peoplespeak, and make it fascinating to the layperson.

Or perhaps the problem runs deeper, to the underlying level of science knowledge and education of the general public. Then again, maybe most people are simply not that interested in science to begin with. Clearly, such speculation could go on and on, and it is pointless. Suffice to say that the following—the discoveries about the moon brought about solely by Apollo—remain recognized by only the smallest percentage of nonspecialists today. Many will seem fascinating mainly because they are being encountered here for the first time.

# The Top Fifteen Amazing Discoveries

1. In preparation for manned exploration, a series of unmanned robotic craft (the American *Surveyor* and the Soviet *Luna* series) soft-landed on the lunar surface in the mid-to-late sixties and sent back clear photographs of their surroundings. The images were shocking: Four centuries of telescopic views from Earth and then space probe fly-bys had convinced astronomers that moon mountains were pointy and sharp. It's how artists had always depicted the moon. The lack of lunar air, water, rain, wind—any sort of erosion—supported these observations. But the actual moon has only rounded mountains, more resembling the Appalachians than the Rockies. Turns out, aeons of countless small meteor impacts have acted like tiny hammers, pounding everything into a totally worn appearance.

2. Almost from the moment that seismometers were planted on the moon's surface and started to beam their findings back to Earth, an astonishing phenomenon came to light. Whenever the moon got clobbered by a hard impact, such as the third-stage Saturn rocket or the abandoned LM deliberately sent crashing to the surface after the astronauts had transferred to the CM, the "moonquake" it produced went on and on and on.

Earth's own ground tremors dampen out and vanish in a minute. On the moon, the shaking just refused to quit. Moonquakes droned on seemingly interminably, lasting for up to two-and a half hours.

The moon rings like a giant gong.

Apparently the moon's interior is mostly solid to the very center, unlike the significant liquid outer core of Earth. The moon's tiny liquid interior (if it exists at all; it is still controversial) is too small to inhibit the spread of earthquake waves through the total 2,160-mile diameter of the moon and back again.

3. The moon's surface is everywhere covered with fine dust, as smooth as baby powder. This regolith proved to be an annoyance: it clung electrostatically to the astronauts' space suits, cameras, equipment, everything. It could not be easily removed—it simply globbed on and stayed there. This dust extends to a depth of at least six feet, and in some areas at least fifty feet deep. It is the sterile soil that the astronauts drove and walked upon, and into which they found themselves sinking a few inches

with each step. Unlike Earth soil, it contains no moisture, no organic materials, no air spaces. It is mostly plain-vanilla oxygen and silicon, $SiO_2$, together with trace elements.

The lunar surface has rocks as well, but these are not like the rocks on Earth. They formed neither from volcanoes (which gave us our igneous rock) nor from any interior solids. Rather, they are all breccias—clumps of the powdery surface material fused together by the heat and pressure of meteor impacts.

The lunar regolith also contains numerous glass beads, also created by the violent heat of meteor impacts as they melted the sandy soil. Indeed, everywhere on the moon one finds the destructive results of micrometeors, small meteors, and large ones. Even the lunar mountains are rounded as if by erosion, though the moon has never felt the gentle pressure of wind, water, or rain. Rather, the relentless poundings of billions of years of impacts have acted like little hammers of Thor to smooth everything as effectively as cascading rivers.

4. On the Fahrenheit temperature scale the moon's night and day range is easy to remember. In most places, the ground reaches 250 degrees by day, and minus 250 by night. This 500-degree range is much wider than any seen on Earth, Mars, or Venus. But it's perfectly in keeping with an airless world. Mercury, closer to the sun, manages an even greater day–night range. Bottom line: any future human colonies must be supplied with prodigious energy sources for both cooling and heating. Imagine a research station that was surrounded by the coldest imaginable Antarctic conditions for two weeks, then by the heat of the Libyan desert for the next two, and then back to Antarctica again. Now add between 100 and 200 degrees of extra cold and heat to both extremes, and that's the moon, every month.

5. Moon rocks are anhydrous, meaning that they contain no water whatsoever. Of course, no one expected to see liquid drops oozing from a moon rock once it was back on Earth and broken open. What was surprising was that, on Earth, many or most minerals are full of hydrogen-bearing compounds, a legacy of the water, moisture, and atmospheric vapor that was the milieu around which many terrestrial rocks formed.

Not so on the moon. You can't even create water using anything you'd find in a moon rock. Moreover, the rocks were also devoid of volatiles (alcohol compounds), nor were there any organic compounds (those that contain carbon).

So, what *is* found in moon rocks? The moon's modest mass gives it just one-sixth of Earth's gravity, not enough to hold onto an atmosphere. If it were endowed with air, then all manner of moderating influences would follow; it would have become a far more fascinating place, perhaps even hospitable to life.

What it *does* contain is oxygen, lots of it, though every bit is locked up in minerals. Based on moon rock analysis and the moon's overall low density (3.34 grams per cubic centimeter, or a little more than three times the density of water), about 45 percent of the moon is composed of a single substance—silicon dioxide, or $SiO_2$. Sandy stuff. Another 30 percent is magnesium oxide (MgO). Three other minerals, FeO, $Al_2O_3$, and CaO—oxides of iron, aluminum, and calcium—are the only compounds with more than 1 percent of the total. Notice that oxygen figures prominently in each, which is why that single element alone accounts for almost half of the moon's mass. If you want to think of the moon in a simple way, picture it as largely a ball of solid oxygen and silicon, with small amounts of a few other elements mixed in.

Earth's crust is also about 45 percent $SiO_2$, though we have far more aluminum and somewhat more calcium than the moon. Still, the similarities with Earth offer important evidence that the moon is essentially a chunk of our own crust, a sort of Adam's rib.

Even more persuasively, the ratio of the three isotopes of oxygen, $^{16}O$, $^{17}O$, and $^{18}O$ (oxygen atoms with varying numbers of neutrons in the nucleus) are almost identical on the Earth and the moon. This is a strong indication that both were once either the same body, or that they formed at the same place in the solar system from the same material. By contrast, meteorites that have come from the asteroid belt or from Mars have very different isotopic fingerprints. All told, the pieces of the puzzle fit sufficiently to reject any notion that the moon originated from somewhere else and was merely captured by our gravity.

A bit more mysteriously, some of the moon rocks exuded a foul smell. Michael Collins recalls being repelled and amazed by the odor when the samples were transferred to the Command Module during the *Apollo 11* flight.

6. Moon rocks were all formed from high-heat events, unlike on Earth. There are three general types: Basalts are the dark lava rocks formed

from cooling lava, which show up to the naked eye as dark blotches that make up the maria (seas). Anorthosites are the "original" moon rocks; they are lightweight and abundant in the lunar highlands. When the moon was a hot, molten magma ocean, anorthosites were the material that floated to the top like pond scum. The final kind of rock are the breccias, clumps of surface material, fused by the heat and pressure of the unrelenting impacts of meteors over billions of years.

There is no sedimentary rock, nothing formed in layers, nothing formed by slowly solidifying wet materials as on Earth, such as limestone, sandstone, or shale.

7. Minerals common to both Earth and moon include feldspar, pyroxine, and olivine.

The substance that is truly of far greater abundance on the moon than on Earth is helium-3. This is helium whose nucleus boasts an extra neutron, and it is quite rare in our terrestrial environment. The significance of this is singular: if scientists ever figure out how to create a practical method of producing electricity from nuclear fusion (and thirty years of work has thus far proved close to futile), then helium-3 would make an excellent fuel—so excellent that it might actually pay to mine it from the moon and transport it back to Earth. It is the single potentially valuable commodity that the moon may offer.

8. All questions about the moon's age have been definitely answered. The moon is ancient. The oldest rocks have virtually the same age as the Earth and solar system—4.6 billion years. The youngest found were 3.2 billion years old, which coincided with the time when volcanic lava flowed most recently on the surface.

9. Earth and moon have a common origin. We know this because, as we've already seen, oxygen has several forms or isotopes. The ratio of normal oxygen, $^{16}O$, with its isotopes, $^{17}O$ and $^{18}O$, acts like a fingerprint, and this identification marker is very similar on the Earth and the moon, proving the two bodies share a common ancestry. Subtle *differences* between lunar and terrestrial oxygen isotopes are now well understood, and allow scientists to determine with certainty whether a meteorite originally came from the moon.

10. Apollo showed us that *all* craters on the moon come from meteor impacts. This in turn has led scientists to believe that this is the case on Earth and other worlds, too. (Only a few odd examples, such as Crater Lake in Oregon, can be found of terrestrial craterlike formations due to actual volcanoes.) Prior to Apollo, many astronomers thought that lunar craters had both volcanic and meteoric origin.

11. The moon is utterly lifeless. This had always been assumed. But after Apollo, we knew it for sure.

12. The moon is leaving us. Thanks to the laser corner cubes left behind at three of the Apollo landing sites, timed pulses from Earth; have been able to determine the moon's distance with one-inch accuracy. This shows that the moon is slowly spiraling away from Earth at the rate of one and a half inches per year. Sometime between 60 and 100 million years from now, it will appear too small to cover the sun, and total solar eclipses will end.

13. The moon's center of mass is not found in its center. Measured by its geometric center, the center of mass is displaced about one mile in the general direction toward Earth. But it is not exactly toward Earth, in fact, the mass center could be found by going from the geographic center in a direction toward the Sea of Tranquillity, where *Apollo 11* landed.

14. Regions of higher density than the moon's average, called mascons, are scattered about the surface, mostly beneath the largest craters or basins. These mascons may either be due to the high density of the impacting body (asteroid or iron-rich meteor) or to concentrations of lava that originally gushed up to near the surface as a response to the impact fracturing. In science fiction, such a mascon was what led researchers to unearth the mysterious monolith beneath the crater Tycho in the film *2001: A Space Odyssey*.

15. Before landing there, scientists feared that the moon's surface might be either a thin crust that would collapse under the weight of a spacecraft, or else be a deep quicksand-like fine powder that a

visitor would sink into, to be swallowed up. Both fears proved groundless. The top few inches of powder are as fine as talcum; beneath that the material is so compact, dense, and free of air spaces it can support any amount of weight. The bearing capacity is similar to a few inches of fine beach sand atop a slab of granite.

And as for that flat-moon business:

The ancient puzzle of why the full moon seems flat, with even lighting across its face, like a disc painted onto the sky, is now solved. If the moon were covered with sheets of rocks, or clouds, or oceans, it would indeed seem three-dimensional and shaded. But its universally powdered surface has the quality of producing innumerable tiny shadows that are removed only when the sun shines straight down, and this happens only at full moon. At this time, sunlight throughout the lunar surface reflects straight back to earthly eyes as if the moon were a movie screen. The powdery surface also explains why the full moon is so dramatically brighter than other phases—ten times brighter than a half moon. At other phases, little unseen shadows on the surface caused by the sun's sideways angle disproportionately dim the incoming light.

All this hardly counted as a wondrous scientific finding. But if the ancient Greeks or Egyptians could have been apprised of this reality, it would have relieved them of a persistent, centuries-long vexing intellectual puzzle that they had no way of solving.

# Origin of the Moon

Doubt *still* remains about the way the moon formed. The 842 pounds of moon rocks, collected in 2,196 separate samples in numerous locations, failed to eliminate competing theories of the moon's genesis, except for one: the moon definitely didn't start out "some other place" in the solar system or universe and get captured by Earth's gravity, as was widely believed in the 1950s.

Although we learned that moon and Earth share an ancestry, this might mean either of two very different things. Choice one is that the two bodies formed from the same cloud of condensing particles (planetesimals); this became the prevailing view among scientists in the years immediately following the Apollo missions. Or, choice two, a much more violent and interesting scenario, is the one explored fully in the next chapter because it is the prevailing view among astronomers today, in the twenty-first century.

# For Nerds: More Apollo Science Findings

Here is a comparison of substances that make up both moon and Earth. Only those that account for at least one-half of 1 percent by weight have been included.

|  | Moon | Earth |
|---|---|---|
| Silicon Dioxide | 45% | 45% |
| Magnesium Oxide | 30% | 7% |
| Ferrous Oxide | 12% | 7% |
| Aluminum Oxides $Al_2O_3$ | 6% | 25% |
| Calcium Oxide | 5% | 16% |

The total quantity of moon rocks returned from the six Apollo landing sites was 841.6 pounds (381.69 kilograms). If one uses the published figure for the final expense of the Apollo program at $24 billion, then moon rocks cost about $30,000 per pound. What they are actually worth in today's marketplace is anyone's guess, since no private citizen is supposed to possess any at all.

The rocks and dust had, by 1975, been broken into 35,600 samples with an average weight of about ten grams—almost half an ounce apiece.

About twelve pounds were given to museums around the world, including Washington's Smithsonian Air and Space Museum, where a chunk is implanted in a theft-proof display that allows visitors to run their fingers over it and actually touch a piece of the moon.

About 95 percent of the lunar material has been kept safe from laboratory analysis (which has already destroyed about three-and-a-half pounds of it—it must be vaporized to be properly analyzed by certain types of spectroscopy).

In addition to the major Apollo returns, the Soviet Union sent two spacecraft—*Luna 16* and *Luna 20*—that successfully scooped up and returned to Russia a total of five ounces (130 grams), all from the Mare Fecunditatus region.

Here are the major science experiments performed on the surface of the moon, and how many of the six Apollo landings set up each experiment:

- Passive seismic, five sites. Measures moonquakes, impacts.
- Active seismic, two sites. Produces known thumps or explosions and measures the subsurface waves.
- Lunar surface magnetometer, two sites. Measures magnetic fields.
- Solar wind spectrometer, two sites, and solar wind composition, five sites. Both analyzed the composition of the particles arriving from the sun.
- Heat flow, three sites. Used holes bored deeply into the ground to measure the way daytime solar heat propagated through the regolith.
- Cosmic ray detector, two sites. Measured the high-energy particles arriving at the moon's surface from places far beyond the solar system.
- Other experiments, most of them set up and left behind by the final Apollo 17: A lunar dust detector, neutron probe, surface gravimeter, atmospheric composition, surface electrical properties, seismic profiling, ultraviolet spectroscope and camera, close-up (microscopic) photography of the surface, charge particle lunar environment, and ion detectors.

# What *Didn't* Apollo Accomplish?

Arguably the biggest disappointment was that the extreme difference in appearance between the moon's near side and far side has not been incontrovertibly answered. Apollo *did* find that the moon's center of mass is located more toward the side facing Earth, and this does suggest that ancient subsurface volcanism rose higher in Earth's direction than toward the other way. This strongly suggests that tidal forces may have influenced greater volcanism on one hemisphere, which in turn created the disparate appearances of the two sides.

# CHAPTER 27

# Back to the Future: Return to the Moon—and Beyond

*. . . And pluck till time and tides are done*
*The silver apples of the moon,*
*The golden apples of the sun.*
—William Butler Yeats, "The Wind Among the Reeds"

A re we really going back to the moon? The exploration has, in a sense, only begun. Although the moon can no longer be honestly called a hot area of research, scientists do harbor unanswered questions and serious ongoing plans for our nearest neighbor, mostly in the design and engineering areas—as a testing place for taking the next great leap beyond.

In a way, the renaissance of lunar interest began in 1998, when the U.S. Army's *Clementine* space probe detected the wispy signature of hydrogen above the moon's poles, which strongly suggested the presence of water ice below. Subsequent studies have confirmed the likelihood that in these large, permanently shadowed polar depressions, like vampires' lairs, may lie hundreds of million of tons of ice. They may take the form of sheets of ice-skating material, or, more probably, be lunar regolith ("soil") mixed with ice like a tossed salad. Either way it is a dramatic change in our understanding of the moon, and a sudden 180-degree shift in future lunar possibilities.

Prior to this finding, the analysis of the 842 pounds of moon rocks brought back by the Apollo mission, and the five ounces delivered to Earth by robotic Russian probes, showed that moon material is anhydrous—utterly without water. Except for a single specimen believed contaminated long ago by a comet that crashed onto the lunar surface, the uniformity of the bone-dry regolith was disheartening.

Humans, of course, need water for survival. At eight pounds per gallon, water is a heavy compound: a small two-by-one-foot aquarium weighs 128 pounds. Because the minimum human need is a gallon a day, even urine-recycling (water extraction) equipment wouldn't be enough: Costly water-laden "space-tankers" would compose a major part of any colonization project. Lunar cities would have to rely exclusively on continuous water supplies from Earth, a prohibitively expensive enterprise.

Such high cost is not balanced by much reward. Nothing commercially valuable exists on the moon—no diamonds, gold, uranium, platinum. Although the moon does have fairly concentrated "ores" of magnesium and lesser ones of aluminum, the cost would greatly exceed that of terrestrial mining. Only the potential prospect of the moon's relatively abundant helium-3 would hold any future commercial value. As explained earlier, helium-3 could provide an excellent fuel for an electricity-producing fusion reactor. It is by no means certain, however, that such a process will ever be feasible. Despite decades of expensive research, no team has yet created a sustainable fusion process that puts out more energy than is required for its operation.

Beyond that, there is very little astronomy or science that needs to be done on the moon since orbiting telescopes and the International Space Station can fully achieve such objectives while floating a thousand times nearer to Earth. Nor does our nearest neighbor have any military value.

It is true that positioning certain observatories farther from Earth's glare and large size offers some advantage over near-Earth orbit. And radiotelescopes could do a nice job on the moon's far side, blocked from all earthly interference. But the down side of such huge expense currently outweighs the potential gain.

Considering all that, it is not very surprising that despite Neil Armstrong's famous "one small step" remark, nearly four decades have passed since humans last stepped boot on the moon, with absolutely no plans for any return until very recently.

Ice changed all that. Robotic space probes have recently been funded, whose purpose is to confirm the existence of frozen water at the poles. One of the most dramatic and exciting is NASA's Lunar Crater Observation and Sensing Satellite (LCROSS) mission, which will be sent crashing into the moon in 2009 to impact the twelve-mile-wide crater Shackleton at the moon's south pole. When the 4,400-pound rocket slams into the surface at 5,600 miles per hour, it will blow away a cloud containing two million pounds of debris, which should rise thirty-five miles above the surface. Fifteen minutes after the impact, the orbiting companion probe will fly through this material and analyze it. If water is detected—then, bingo. If LCROSS reveals good news, perhaps the weighty requirement of water-hauling can largely vanish, and future moon colonies suddenly become infinitely more practical. In addition to providing for human biological needs, water's elements can be readily separated or disassociated: the plentiful energy from the sun (or plutonium-driven RTG electricity-makers) can extract hydrogen and oxygen from lunar ice to yield the most efficient possible rocket fuel, one already employed as the exclusive propellant for the Space Shuttle's main engines.

Since the moon's escape velocity is merely 5,000 miles per hour, it's not just a science-fiction scenario that envisions moon colonies as jumping-off sites for more distant destinations such as Mars. Instead of hauling heavy fuel against Earth's 25,000-mph escape velocity, rockets can tank up on a moon base with a fresh supply of hydrogen and oxygen.

No inherent practical problems prevent such an eventual venture, and the moon's extreme daytime heat (250 degrees) could be mitigated by simple shading. Also, air-conditioning could be run using solar panels for electricity. The moon's minus-240-degree nighttime, in a span that lasts for fifteen days, is more of a problem, and solar power wouldn't then be available. Here, the only solutions would either be electricity generated by the hydrogen and oxygen from lunar polar ice (if it exists) or, more probably, power supplied by adequate numbers of plutonium power plants, the reliable old, tried-and-tested Radioisotope Thermoelectric Generators. More likely, a permanent moon colony would be placed at or near the poles, where the sun is always low in the lunar sky and the ground doesn't heat up as much, and where that putative ice would lurk nearby.

President George W. Bush's surprise 2004 announcement of intentions for manned missions to resume to the moon has changed the U.S. space agency's priorities. Already, NASA has switched its focus from the International Space Station (after 2013), and the soon-to-be-abandoned Space Shuttles needed to service the station, to the goal of returning to the moon, establishing semipermanent bases there, and eventually moving on to manned missions to Mars. These present formidable challenges. The unfortunate fact is that many of the dies and equipment for the Saturn V that was employed for the Apollo missions have not been saved. A new rocket must start almost from scratch, as, perhaps, it would have anyway.

Some have suggested that the International Space Station (ISS) be used as a jumping-off point for manned lunar missions, but that simply would not work. The ISS was deliberately placed into a high-inclination orbit so that it could be reached from launch sites in both Russia and the United States. The moon's orbit, by contrast, follows the sun–Earth zodiacal plane that is much more closely aligned with the equator, and could not be practically accessed from the ISS. Instead, a brand-new space vehicle is being envisioned for the project. The argument for establishing a base on the moon is simple: If humans are to live on much-more-distant Mars, then the technologies and bubble-encased home environment can and should be tested on the moon first, where people could much more easily return home if something goes awry. Gary Martin, NASA's long-term strategic planner, who is essentially that agency's "space architect," told the author in 2005 that the moon's surface could provide important "test-bed" activities toward a later Martian visit, a place to try out space suits and mechanical rovers that would need to negotiate sandy soil, and other such equipment.

Buzz Aldrin enthusiastically supports this thinking. Although he feels that the moon is too harsh to justify any sort of permanent settlement, it should be used as a stepping stone to Mars—a goal he is actively pursuing. But not, he says, for simply brief manned visits to the Red Planet, the way we did with the moon.

"Permanence is needed on Mars," he told the author in 2006, "not just one or two visits." The former astronaut argues that there be five-year tours on Mars, because if the astronauts there depart during the once-every-twenty-six-month orbital opportunities to return to Earth, the equipment and colony would simply be abandoned until the next arrivals. Instead, some people—perhaps even space tourists—should each stay there through at least two of the biennial return "windows."

The down side to the use-the-moon-to-reach-Mars plan is that it's easier, energy-wise, to launch directly for Mars from Earth than have to slow down and stop at the moon, and then escape the moon's gravity for the next leg of the trip. By this reasoning, the moon should merely be used in the test-phase of the Martian program. Critics also argue that the vast expense for such a program is already pulling funds away from the highly successful and much less costly unmanned robotic missions such as the Cassini craft to Saturn, that have already proved themselves capable of enormous scientific returns with no danger to humans. Planners will also have to see if the next generation of Americans is excited by all this, or if there's a "been there, done that" mentality that bogs down everything in bureaucratic epoxy. "It doesn't make sense to send manned missions anywhere right now," says NASA-Ames eclipse expert Fred Espenak. "We should wait until much later in the century, when the technology makes it more feasible. But these programs are always driven more by politics than by science."

Geoffrey Marcy, the renowned University of California extra-solar planet-finder, lamented in 2006 that funds for finding life elsewhere in the universe were being slashed by NASA to make room for the moon missions. He described his colleagues as "deeply depressed," and wondered "if people are truly inspired by going somewhere we've already been." A 2005 *New York Times* op-ed piece was even harsher:

> [The proposed missions to the moon] are expensive and provide too little return. But politicians know that the American public identifies progress in space with human astronauts. The Bush administration's solution is to create an impossibly expensive and pointless program for some other administration to cancel, thus bearing the blame for ending human space exploration. The return to the moon is not a noble quest. It is a poison pill."

Indeed, few in and out of NASA privately dispute that the huge costs for a future moon-and-then-Mars mission make it a problematic exercise. As of 2006, the official policy was "pay as you go," which meant that NASA would have to fund it out of their already-existing $16-billion-a-year budget. This is clearly a different mind-set from the heady Apollo years, when we were to beat the Russians at all costs.

Then, too, one can easily remember how Richard Nixon was no great fan of Apollo, since it was the brainchild of his predecessor, John F. Kennedy. The current timetable calls for the soonest moon landing to occur in 2018, with 2020 starting to be increasingly quoted. This means that three future administrations will have to fund it. In such situations, the usual outcome is for half steps and partial funding, which would greatly drag it out further. Moreover, even the 2020 timetable assumes that each of the interim steps will be achieved on time, which few analysts in or out of the government believe for one second.

Once back at the moon, the plan calls for evaluation, study, and design based on the new moon findings in order to then construct a manned mission to Mars. So for those who will not be excited by a "been there, done that" moon destination but would indeed like to see humans visit Mars, the best advice is to start believing in reincarnation. When Apollo concluded in 1972, most media commentators believed we'd be on Mars during the 1980s—that is, within fifteen years or so. For decades afterward, the vague dream always had it that Mars was "fifteen or twenty years away." Now, during the first decade of the new century, realists do not expect Americans on Mars until at least 2030, if then. In short, its realization—like the ever-lengthening hallway of a dream—keeps moving farther into the future.

Right now the betting is that it is more likely that the Chinese, flush with cash and a desire to show the world that it is a first-rate technological power, would be willing for prestige's sake to fund the Moon and Mars. The first space food eaten on Mars, goes this reasoning, is more likely to be chow fun than beef jerky.

Of course, no one can predict which way things will unfold. Currently the United States is the only country with publicly announced specific plans to return humans to the moon, complete with a crude timetable. Rather than paraphrasing, here is the January 14, 2004, White House announcement, presented verbatim:

*The United States will begin developing a new manned exploration vehicle to explore beyond our orbit to other worlds—the first of its kind since the Apollo Command Module. The new spacecraft, the Crew Exploration Vehicle, will be developed and tested by 2008 and will conduct its first manned mission no later than 2014. The Crew Exploration Vehicle will also be capable of transporting astronauts and scientists to the International Space Station after the Shuttle is retired.*

*America will return to the moon as early as 2015 and no later than 2020 and use it as a stepping-stone for more ambitious missions. A series of robotic missions to the moon, similar to the Spirit Rover that is sending remarkable images back to Earth from Mars, will explore the lunar surface beginning no later than 2008 to research and prepare for future human exploration. Using the Crew Exploration Vehicle, humans will conduct extended lunar missions as early as 2015, with the goal of living and working there for increasingly extended periods.*

*The extended human presence on the Moon will enable astronauts to develop new technologies and harness the Moon's abundant resources to* *allow manned exploration of more challenging environments. An extended human presence on the Moon could reduce the costs of further exploration, since lunar-based spacecraft could escape the Moon's lower gravity using less energy at less cost than earth-based vehicles. The experience and knowledge gained on the Moon will serve as a foundation for human missions beyond the Moon, beginning with Mars.\**

Those are the words, the dates, the intentions of the second Bush presidency.

But by September 2005, a mere twenty months later, the official NASA timetable to the moon had already been pushed back to 2018 at the earliest. It had fallen behind three years in less than two. By 2007 the date had morphed to 2020. The future kept receding relentlessly.

On the plus side, instead of mere intentions, here was a concrete plan, complete with rough designs and bidding. Moreover, we now had a timetable. It went like this:

First comes development of the CEV, the Crew Excursion Vehicle, the replacement for the Shuttle. Aboard it, astronauts would ride above and in front of the rockets that vault it into space, instead of hazardously alongside, the way the current Shuttles operate. It will also have escape rockets capable of pulling astronauts to safety in case of an accident either on the launchpad or already in flight. Because of this, NASA estimates that it will be ten times safer than the Shuttles, with a projected failure rate of one in 1,000 to 2,000 flights, rather than one in 100 to 200, today's odds.

This vehicle could carry up to six astronauts to the Space Station. For moon missions, it would rendezvous with moon-ready components already in orbit, that had been brought up by a new giant rocket capable of placing 125 tons in Earth orbit. The CEV would come in two versions, Block I and Block II, just as the Apollo capsules had, with each designed for either near-Earth servicing of the ISS, or to take astronauts much farther, to the moon and beyond.

As presently envisioned, moon missions would entail the new rocket hurling a smaller rocket into orbit, which the crew, arriving above Earth on the CEV, would then attach to use as propulsion to the moon. In short, it's back to the old "Earth-orbit rendezvous" idea that Wehrner von Braun and his friends very nearly used in the first place. Then, once at the

\* From a statement made by President George W. Bush on January 14, 2004.

moon, the mission would proceed pretty much as the Apollos had, with the unneeded descent stage abandoned on the moon when the astronauts left to rendezvous with their waiting return-ship in lunar orbit. It all seems so familiar.

Maybe too familiar. The public quickly tired of the original moon missions, which lost all live television and front-page coverage after the first astronauts had accomplished their goals. Can we realistically expect that a reprise would fare any better in the public mind? Everyone was excited when Hillary and Norgay first scaled Everest; who paid attention to the next men who reached its summit? Some wonder whether this is a project that resembles the International Space Station—an expensive, hazardous enterprise that gets scant media coverage, and with which the world is thoroughly bored.

By 2006 the entire return-to-the-moon project was acquiring its own nomenclature. In August NASA announced that the new CEV would officially be called *Orion*. The Crew Launch Vehicle, that new giant rocket with a familiar mating of segmented solid-fueled boosters and bulbous liquid fuel tank, was expected to be named *Ares I* when set up to carry astronauts, and *Ares V* in the cargo-only configuration. The new Lunar Surface Access Module, already abbreviated LSAM, was to be named *Artemis*, though, unlike *Orion*, this was not yet cast in stone.

As for space goals, servicing ISS and going to the moon was a "given." But now, further manned test-flights to places that did not have a strong gravity and therefore a fuel-intensive escape requirement were starting to appear in the plans, prior to Mars and maybe even prior to the moon landing. A round-trip to the Lagrangian-2 point 900,000 miles away, where Earth's gravity is balanced like a hovering seesaw with the far greater pull of the sun. Then, perhaps, a trip to a "near-Earth object"—the NEO would probably be a passing asteroid. Its minuscule gravity could let astronauts hop down using lightweight equipment, and then depart with very little thrust.

Grand plans, ambitious steps that continue to gel as this volume goes to press.

We could leave it at that, and perhaps let the optimism of space travel to other worlds be our last word. Ardent supporters of manned exploration and even future colonization of other planets and moons, such as the members of The Planetary Society, are keeping fingers crossed that these steps do get fully funded—as is indeed possible.

But that would not be entirely honest. While government projects, with their promised profits to aerospace companies, do indeed acquire a slow if inexorable lavalike inertia that is hard to stop, the widespread whispered wisdom in the science community is that returning to the moon is a poor way to reach Mars, and an expensive diversion of space dollars. Because such a view is so widespread, it is hard to predict whether an American moon return will actually ever come to fruition, the present timetable notwithstanding.

Comparisons with Apollo do not inspire optimism. The initial moon race had the NASA beast continuously whipped like an ox by the threat that the Russians would get there first. No such competition exists today. Then, too, Apollo was budgeted by the "whatever it takes" philosophy, while today's program is not. Moreover, Apollo was unique, with Congress and the public fully behind it, and was noncancelable; the current situation, as we've seen, requires the cooperation of three additional administrations and sets of elected officials. The latter may well realize that their constituents would not be particularly upset if a moon replay were to be delayed indefinitely. All told, conditions are very different today.

One should also be cautious about the naming of the various planned craft, and expect numerous redesignations before the mid-2010s. Just as the Hubble Space Telescope's replacement, the Next Generation Space Telescope (NGST), has become the *James Webb*, and the Freedom Space Station morphed into the ISS, one cannot count on any labels until they are actually painted on the ship. Odds are, *Artemis* and *Ares* will be rebaptized between now and then.

# Mars and Beyond

Mars beckons—by default.

It's not that Mars is such a friendly planet. Quite the contrary. Living there would be far more harsh and difficult than taking up residence atop Mount Everest or deep under the sea. Mars offers no breathable air, and the typical temperature is eight below zero. The Red Planet's lack of a magnetic field means no shielding from solar or cosmic radiation. It is almost unimaginably inhospitable. The only reason utopians and manned-space planners think "Mars" is that there is simply nowhere else to go. Venus is a pressure-cooker with temperatures that far exceed those of an oven. Mercury has an impossible 1,000-degree day–night variation, and no air whatsoever. The planets from Jupiter outward have no surfaces on which to land. That leaves Mars by default. Like vacationers with money who face a boarded-up travel agency, the notions of colonizing Mars are born not of romance but of desperation. It's not hard to choose a realistic destination when there's only one in the solar system.

Having a few of us spend time there—and eventually establish a long-term bubble-enclosed colony of expatriate earthlings—is such an established staple of sci-fi and science, it wouldn't take much of a "public sell" to fund the steps to the Red Planet.

Rather than rehash all the obvious if expensive components (four-to-six-month travel time, equipment shipped ahead and awaiting the arriving astronauts, and so on), it might be good to look at two lesser-discussed possibilities.

The first involves cost. It's much more than twice as difficult to get there and return, than to just get there. So a nation willing to be bold could leapfrog ahead of everyone else, and shave a decade off the development time, simply by sending one-way volunteers to land where years of supplies had been previously parachuted by robotic cargo drops. These Marsonauts would assemble a protective station from ready-to-fit modular components—everything they'd need for years of Martian living. As the further components and rockets were built back home, the colonists would be brought their ride home in four or six years. Finding astronaut volunteers? Never a problem. By this method, a dedicated program could easily land people on Mars by the early 2020s.

But here's the rub. There is increasing evidence of great biological hazard beyond Earth's protective cocoon. It is one thing to venture up more than 60 miles to leave our planet's life-giving atmosphere, and to orbit at a height of 250 miles or so. Beyond a few thousand miles, however, outbound astronauts also leave the magnetosphere, the barrier against incoming radiation from the rest of the cosmos. The dangers of spending months or years outside of this shield, especially if solar coronal mass ejections (CMEs) throw deadly subatomic pellets at the departing craft, are not trivial. A small shielded safe room on the Mars-bound craft, and one on the moon and on Mars, might protect against the most intense CMEs. But astronauts cannot hide there permanently, and ongoing charged particles and ionizing radiation trickle in continuously.

In 2006, astronaut Shannon Lucid, the woman who has spent the most time (223 days) in space, told a *Discover* magazine interviewer: "Radiation could be a showstopper."

Indeed, a mere month prior to the *Apollo 17* final mission to the moon, an intense CME sent a blast of solar radiation sweeping past Earth and moon. Had the astronauts been on the lunar surface at the time—or even aboard their inbound or outbound craft—they would have received a fatal radiation dose. It was one more reason that NASA officials were quietly glad that Apollo's final three missions had been canceled. Quit while you're ahead, before someone gets killed out there. . . .

Then, too, continued habitation aboard the ISS has shown that extended weightlessness is not salutary for health. Permanent bone and muscle loss on the average of 1 percent per month is suffered by everyone who spends more than a few weeks in space, and recent studies have shown shrinking heart capacities and other ills.

The Martian environment is more temperate than that of the moon, but it still allows full entry to the surface of high-speed charged particles from the sun, plus X-rays and gamma-rays from space. Even if cancers do not rapidly develop, some astrobiologists estimate that the ongoing radiation will damage or destroy as much as 40 percent of the brain in a mere two years on the Martian surface and en route. Such losses—far exceeding the 5 percent annual neuron necrosis suffered by many Alzheimer's patients—would be incapacitating. Brief intense radiation bursts could produce acute radiation sickness, including vomiting, which would be very serious within a space suit.

Antibiotics and other medicines invariably deteriorate, too, losing their potency, and advanced treatment options such as CAT scans and complex surgery would be unavailable for a seriously ailing space-farer. Long-term space travel is obviously not for hypochondriacs or jocks who couldn't stand to see their beloved bodies deteriorate.

In short, it's entirely possible that habitation of Mars, even for a few short years, may simply be too debilitating to anyone's health. One can only imagine the public impact of the announcement of the death in space or on Mars of one or more of its astronaut heroes. If the hazards prove to be as formidable as many experts fear, would the project even get the go-ahead?

Balancing this, of course, is the collective desire to see the chocolate-red world for ourselves. And with Mars, perhaps the public's fancy can be held for more than the standard fifteen minutes. One possible way to keep the public excited about manned space travel may be a suggestion made by Buzz Aldrin, that when the new heavy-lifting rockets have been built by 2012 or so, space tourism is actively encouraged. And not just for the wealthy, who could afford to shell out huge amounts for the experience. "We should have lottery-like selections that allow a national selection process, that lets ordinary people experience space travel."

# The Far Future

Meanwhile, those of us forever confined to terra firma will continue to explore space, and especially the moon, on our own. It's more than vicarious travel: the advent of inexpensive optics and computer and motor-driven telescopes have put high-quality equipment within the reach of almost everyone. But it's not likely that human space travel will ever be discernible to those gazing up at the night sky. Even the best telescopes have a resolution no better than two city blocks on the moon, so individual lunar spacecraft will never be viewable from here, just as there remains no visible trace of our previous landings through terrestrial instruments. If a large domed bubble is ever erected to house lunar colonists, and if it does exceed two blocks in scale, then it might be marginally glimpsed as a speck on the riotously detailed lunar terrain by a Hubble-class telescope. But since the naked eye cannot detect lunar features smaller than 300 miles across, and no man-made Earth structure is that large except for the Great Wall of China, it is not likely that earthbound humans will *ever* gaze on a moon transformed from the way it looks today, from the way it has always appeared.

Yet *it* will eventually transform itself. Spiraling away from us at the rate of one-and-a-half inches per year, the moon will continue to appear smaller as time goes by. Right now its disc matches that of the sun, but in 100 million years the two will seem noticeably different in size. Because tidal strength varies not with the square but with the cube of distance, our oceans' present three-foot average rise and fall will diminish, and become much less animated than in our current age.

Eventually, in about four billion years, the moon will stop its outward journey and maintain a steady forty-day revolution period, matching Earth's newly lethargic spin. The two bodies will face each other without blinking, eye-to-eye, as Pluto and its moon, Charon, do today. One hemisphere of Earth will see the moon throughout every night; real-estate properties on the other side will glimpse no trace of it. Would this influence land values?

Then, after a few million years of stability, the moon will slowly begin moving the other way, returning like a salmon toward its place of birth. In that remote future era when humans and all other terrestrial life will be gone, today's three-foot average rise and fall of the seas will have morphed into an unimaginable daily range of more than 12,000 feet. More likely, however, there would be no oceans in that future epoch since the sun is expected to continue its rise in luminosity and bake Earth into a sterile 700-degree near-Venus-like hell in about 1.1 billion years. In that case, only the air and the ground itself would distort from the menacing nearby moon. The present eight-inch "solid tide" would grow to a daily surface up-and-down half-mile contortion.

Finally the moon will venture within the Earth's "Roche limit" of 10,000 miles, where it will be torn apart by tidal forces from our gravitational field. In this, its final act, the moon must then turn into an orbiting ring of countless particles. Any future descendants, perhaps survivors of the "hot sun" and then "large red giant" phases who now return to a freshly rehabitable Earth illuminated by a smaller white dwarf sun, will gaze upward into a moonless sky. Perhaps they will still have genetic memories or distant records of the time, long ago, when a moon once danced across the night, and when people first trod its powdery soil.

But replacing the romance of the full moon will be an even more spectacular Saturn-like ring. The moon will provide something beautiful to see in the nightly heavens even then.

For now, and in the immediate centuries to come, the real target for earthbound moon explorers will continue to be its fascinating and endless detail. Using a mere six-inch telescope, the renowned lunar cartographer Julius Schmidt published a map in 1878 that showed 32,856 craters on the moon. Today's backyard telescopes can reveal half again as many, underscoring that observers can never run out of things to see on our nearest neighbor, especially since changing shadows caused by the ever-moving sunrise line will eternally create new lighting effects. And astronauts, should they return there, will similarly never run out of places to probe, as they possibly prepare for their next biggest adventure.

The story is never over.

# Bibliography

Baker, David. *The History of Manned Space Flight*. Crown Publishers, 1982.

Collins, Michael. *Carrying the Fire*. Farrar, Straus and Giroux, 1974.

Compton, David W. *Where No Man Has Gone Before: A History of Apollo Lunar Exploration Missions*. U.S. Government Printing Office, NASA SP-4214, 1989.

Cortright, Edgar M., ed. *Apollo Expeditions to the Moon*. U.S. Government Printing Office, NASA SP-350, 1975.

MacKinnon, Douglas, and Joseph Baldanza. *Footprints: The 12 Men Who Walked on the Moon Reflect on Their Flights, Their Lives, and the Future*. Acropolis Books, 1989.

McAleer, Neil. *The Omni Space Almanac*. Pharos Books, 1987.

Reynolds, David West. *Apollo: The Epic Journey to the Moon*. Tehabi Books, 2002.

Turnill, Reginald. *The Moon Landings*. Cambridge, 2003.

U.S. News and World Report. *U.S. on the Moon*. U.S. News and World Report, 1969.

Wylie, Francis E. *Tides and the Pull of the Moon*. Berkley, 1980.

# Index

*Boldface page numbers indicate photograph inserts.*